A COURSE IN WORLD GEOGRAPH

Book 4

THE BRITISH ISLES

Physical and Regional

By

J. H. LOWRY, M.A., B.Sc.(Econ.)

Senior Geography Master
Cranleigh School

Fifth Edition

MAPS AND DIAGRAMS BY J. H. LOWRY
AND CARTOGRAPHIC ENTERPRISES
ILLUSTRATIONS BY KLAUS MEYER

EDWARD ARNOLD

© E. W. Young and J. H. Lowry 1979

First published 1960
by Edward Arnold (Publishers) Ltd.,
41 Bedford Square, London WC1B 3DQ

Reprinted 1961 (twice), 1962, 1964
Second Edition 1964
Reprinted 1965, 1967 (twice)
Third Edition 1968
Reprinted 1969, 1970 (twice), 1971, 1972 (twice)
Fourth Edition 1973
Reprinted 1974, 1975, 1977, 1978
Fifth Edition 1979
Reprinted 1981

British Library Cataloguing in Publication Data
Young, Eric William
A course in world geography.
Book 4: The British Isles—5th ed.
1. Geography—Text-books—1945–
I. Title   II. Lowry, John Henry
910   G128
ISBN 0-7131-0396-5

The full course comprises:

BOOK 1 — People in Britain

BOOK 2 — People Round the World

BOOK 3 — Regions of the World

BOOK 4 — The British Isles

BOOK 5 — The World: Physical and Human

BOOK 6 — Europe and the Soviet Union

BOOK 7 — North America

BOOK 8 — East Africa

BOOK 9 — Central Africa

BOOK 10 — The World: A Systematic Geography

Printed in Great Britain by
Butler & Tanner Ltd, Frome and London

FOREWORD

THIS BOOK forms one of a series designed primarily for students who will take geography at GCE 'O' Level. Experience shows that, used in conjunction with more advanced texts, it can also form a useful basis for VI Form studies.

In keeping with other books in the series formal statement of geographical fact has been cut to a minimum, and the necessary data are expressed very largely in the form of maps, diagrams and illustrations. The responsibility for their interpretation is laid expressly on the student, with or without the teacher's help according to circumstances. In this way it is hoped to encourage the development of genuine geographical understanding and interest, and to discourage mere memorisation. The layout of every page has been carefully devised so that, although each book is written in chapter form, the material does in fact fall into units of one, two or at most four pages. Lesson-planning will be found to be much facilitated by this arrangement.

Books 1, 2, 5, 7, 8 and 9 are by E. W. Young, and Books 3, 4 and 6 by J. H. Lowry in association with E. W. Young, who has planned the series as a whole.

Book 4 falls into two distinct though interrelated parts. Chapters 1–3 cover the basic physical geography required for the G.C.E. Ordinary Level examination. Chapters 4–21 deal with the geography of the British Isles on a regional basis, also at G.C.E. Ordinary Level standard. The physical section is illustrated, wherever possible, by reference to examples from within the British Isles, and there are frequent cross-references between the physical and regional sections of the book. The final two chapters provide a thorough revision. The author has assumed that an atlas will be available for each pupil, but a large geological map of the British Isles has been included (pp. 64–65) in addition to the maps and diagrams for each topic and district.

The author acknowledges with thanks the help given him by representatives of the industries and organizations mentioned in the text. Special thanks are due to Mr. J. Spurway, B.Sc.(Econ.), Senior Geography Master of Falmouth Grammar School, who read the proofs and offered many valuable suggestions.

J. H. L.

FOREWORD TO THE FIFTH EDITION

The geography of the British Isles is undergoing profound and rapid changes. Such developments as the exploitation of North Sea oil and gas, new techniques in steel production, the spread of motorways, expansion in petro-chemicals, inner city decay, the re-evaluation of the role of coalmining and entry into the Common Market are affecting virtually the entire pattern of industrial and agricultural production. This new edition examines all these trends, both systematically and in their regional implications.

Users of previous editions will thus find substantial alterations in most chapters, but the basic plan and aim of the book remain essentially the same, namely to present geographical concepts in a clear, intelligible and relevant way.

SPECIAL NOTE

Throughout this book certain words are printed in heavy type, thus : **escarpment**. It is suggested that these words should be collected, chapter by chapter ; that a satisfactory meaning should be attached to them ; and that the words and their definitions should be recorded in the pupil's own " geography dictionary " or in any other form that suits the teacher's particular approach. By this means a sharper edge may perhaps be left on terms which all too often are blurred by generalisation.

CONTENTS

5

CHAPTER 1

Rocks and Mountains

DESOLATE MOORLANDS ; precipitous sea-cliffs ; rich agricultural plains ; giant industrial cities ; great bogs and undrained marshes ; land reclaimed from below sea-level . . . Where in the British Isles can each of these be found ? How well do you know your country ?

Which sea-side resorts are likely to get most rain in summer ? Where and when can you go ski-ing in Britain ? Where are the best places for sea-fishing, swimming and boating ?

When you leave school and have to choose a job, will you know which occupations offer the best prospects, which industries are gradually declining, and which are expanding and becoming more prosperous ?

You may know the answers to all these questions, but can you explain *why* there is such a great variety of scenery, occupations, weather conditions and ways of life in such a small place as the British Isles ? *Why* do hops grow best in Kent and Sussex, whilst west Cornwall is more suitable for spring flowers and vegetables ? *Why*, in January, is it as warm on the coast of northern Scotland as in the Isle of Wight ? *Why* are there such broad stretches of barren land in central Wales while people are tightly packed in cities on the South Wales coast ? *Why* has London continued to grow until it contains over 7 million inhabitants ? *Why* have most canals in Britain fallen into decay ? *Why* does there happen to be a town, or village, or city where your school is situated ? In the course of this book we shall be seeking the answers to these and to a multitude of similar questions.

The scenery of any particular place, and the occupations and livelihood of its inhabitants, are closely related to its geography ; the fertility of the soil, the amount of sunshine and rainfall, the number of roads and railways, the presence or absence of mineral deposits such as coal or iron ore, and so on. In trying to understand and explain what we see on the surface of the earth, we also use knowledge of the earth's interior. Miners find that the deeper they dig into the earth the higher the temperature at

6

the bottom of the shaft. Scientists tell us that the heat becomes so intense with increasing depth that a point is reached, some 40 to 50 kilometres below the earth's surface, where all known substances would melt.

At one time in the distant past the earth was probably a whirling mass of flaming gases. Throughout millions of years these gases gradually cooled, and many of them condensed into liquids. Later still, a thick scum began to collect on the cooler surface of the liquid, like the layer that forms on top of hot fat when it is poured into a dish from a frying-pan. Very slowly this layer solidified into a number of rigid 'plates'* up to 65 km thick, forming the Earth's *crust*. They float on a deeper layer of semi-molten material (*magma*)

Section through the Earth

CRUST
mainly solid
c 65km thick

CORE
solid or
semi-solid
c 6000 km
diameter

MANTLE
semi-solid?
c 3000 km
thick

*For greater detail
see Book 10 p. 16*

which becomes fully molten wherever the pressure of the outer crust is relaxed. Local pressure changes, caused by the movement of 'plates', allow magma to escape either at ground level or on the sea bed or into the outer layers of the crust. There, in time, it cools to form part of the crust itself.

The materials forming the continents are parts of the solid outer crust, and the ocean waters fill up hollows and basins in its surface. The substances making up the crust are known as **rocks.** We may forget that these rocks are there, close beneath our feet, because they are usually covered by soil, crops, forests, cities and roads; but in places like railway cuttings, quarries and sea-cliffs we can see them quite easily.

To most people the word 'rock' means something very hard and solid, but in geography it is used to describe *any* substance forming part of the earth's crust. Thus soft mud and sand, or crumbly chalk, is as much a rock as hard granite. There are scores of different kinds of rock, but each belongs to one of three groups: **igneous, sedimentary** and **metamorphic.**

* Book 10, pp. 16–29.   7

1. **Igneous** rocks have been formed from the molten magma beneath the solid crust. They get their name from the Latin *ignis*—' fire '. Sometimes great quantities of magma are forced up through splits in the crust, either to the earth's surface or just below it. Here they cool and turn solid, usually becoming extremely hard. Examples of such rocks are granite and basalt. Igneous rocks are widespread in the British Isles, great masses of them occurring in such places as Devon and Cornwall, Scotland, and parts of Northern Ireland. Because of their hardness they can withstand for millions of years the attacks of the winds, waves, rain and storms which constantly assail them. Even so, all rocks are ultimately destroyed by the forces of nature. The small islands in this photograph are all that remain of land which, long ago, formed part of a seaward extension of West Cornwall. (*See also page 203.*)

In the laboratory you have probably watched the cooling of hot solutions such as copper sulphate and sodium carbonate. At a certain temperature, crystals appear. If the cooling is rapid, many tiny crystals are formed ; on slower cooling, the crystals are bigger. Many igneous rocks are crystalline. In some cases, where the molten material ran out on to the earth's surface and cooling was quick, the crystals of rock are so minute that one needs a microscope to see them ; but when the cooling took place below ground it was a slow process in which large crystals were sometimes formed. This sketch shows actual-size crystals of igneous quartz. Much of the sand of the sea-shore is made up of small particles of this material.

Now look at the geological map of the British Isles on pages 64–65. It shows what kinds of rock form the earth's crust in the various regions of Britain. Notice carefully the places where igneous rocks (labelled ' volcanic ' or ' granite ', etc.) are indicated. Look next at the relief map of Britain in your atlas. (*1*) Can you see any link between the geological and relief maps ? If so, what is it, and how do you explain it ?

The rocky coastline at Land's End, in Cornwall. In spite of endless buffetings by Atlantic waves, the granite cliffs here stand out boldly into the ocean.

(2) *Make a tracing of this map and add arrows to show where the less resistant shale is being destroyed by the attacks of the waves.*

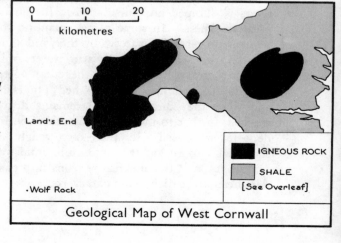

0   10   20
kilometres

Land's End

· Wolf Rock

■ IGNEOUS ROCK
▨ SHALE
[See Overleaf]

Geological Map of West Cornwall

After the flood ... MUD
Cleaning up in an Oxford street after the River Thames had overflowed its banks.

2. **Sedimentary rocks.** On their journey to the sea most rivers carry a great quantity of mud. Most of it is soil and tiny fragments of rock which have been washed into the rivers, and it is called **alluvium**. The River Thames carries 85 000 tonnes of alluvium into the North Sea every year: the Mississippi carries the fantastic figure of two million tonnes of rock-waste seawards *every day*. The waters of the Gulf of Mexico are discoloured far out to sea by alluvium from the Mississippi.

During millions of years this river-borne silt from the world's rivers is dumped on the sea-floor. In time it accumulates to a great thickness. In some places quantities of sand from desert regions are carried out to sea by the wind and eventually come to rest on the sea-floor. In warm waters millions of tiny sea-creatures thrive. Some, like the coral polyps shown opposite, grow in huge colonies on the sea-bed; others float near the surface, and when they die their shell remains drift downwards to the sea-bed, further to increase the depth of **sediments**, as all these deposits are called. By their own weight these sediments are squeezed into rocks; the muds into mudstones, the sands into sandstones, and the shellfish remains into chalks and limestones. All these are called sedimentary rocks.

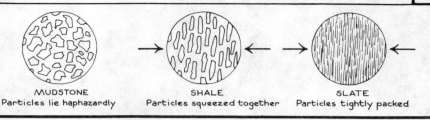

| MUDSTONE | SHALE | SLATE |
| Particles lie haphazardly | Particles squeezed together | Particles tightly packed |

3. **Metamorphic rocks.** After they have been formed, both igneous and sedimentary rocks are sometimes tightly compressed by movements in the earth's crust which we shall consider overleaf. Sometimes, too, they are affected by intense heat, as fresh masses of molten magma force their way towards the surface.

Changes brought about in the rocks by such heat and pressure are sometimes so great that new kinds of rock are formed. For example, a crumbly mudstone may be changed, first into a shale, and later into slate, becoming harder and more compact in the process. Similarly dull, coarse limestone is sometimes changed into tough, gleaming marble, sandstone into quartzite, and granite into gneiss. In most cases the newly formed metamorphic rock (from *meta*—Greek for 'change') is quite hard and resistant; in fact, very like an igneous rock.

Folded strata on the Dyfed coast,
South Wales.

In parts of Britain sea-shells are found embedded in sandstones
many hundreds of feet above sea-level.   In mountainous regions
like the Alps similar shells are found on peaks thousands of
metres high.   To understand how they got there you must realise
that the earth's crust on which we live is not absolutely rigid—it
is constantly warping and bending as enormous forces from
beneath cause stresses and strains.   In places these forces have
lifted sediments bodily upwards to form dry land—the **strata**
(as compressed layers of sediment are called) remaining roughly
level.   Elsewhere, what was once dry land has disappeared
beneath the waters.   Britain, for example, was at one time joined
to the Continent where the English Channel now separates
Dover and Calais.

The uplift of sediments is not always a simple process.   Strata
are often warped and twisted into upfolds (**anticlines**) and down-
folds (**synclines**) as in the diagram. **Folding** such as this will
occur when sediments are squeezed by the movement together of
harder, more resistant beds of rock.   Notice in the diagrams how
the folding becomes more complicated as crushing continues.
*Remember that these movements, as we measure time, take place
extremely slowly.   A shift of 1 centimetre in a year would be very
rapid.*   Nevertheless, continued through millions of years these
movements are responsible for the creation of gigantic mountain
ranges like the Alps, the Himalayas and the Rockies.   Mountains
built in this way are called **folded mountains.**

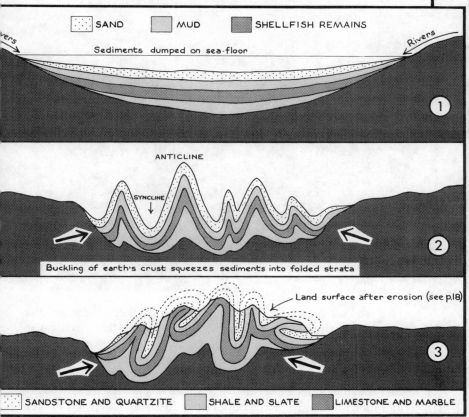

SAND   **MUD**   **SHELLFISH REMAINS**

Rivers   Rivers

Sediments dumped on sea-floor

①

ANTICLINE

SYNCLINE

②

Buckling of earth's crust squeezes sediments into folded strata

Land surface after erosion (see p.18)

③

SANDSTONE AND QUARTZITE   **SHALE AND SLATE**   **LIMESTONE AND MARBLE**

During a hot summer ponds often dry right up, and the mud on the bottom receives the full strength of the sun. As the mud dries out it cakes and cracks into patterns like those in this picture.

The cracks are caused as the mud shrinks in drying. Sediments raised from below sea-level will gradually dry in the same way. The photograph overleaf shows cracks and splits in limestone in the Pennines. (See also pages 110 and 170.) The larger splits, called **joints,** frequently break up the rock into great blocks roughly rectangular in shape, and this helps in quarrying the material.

The Great Glen, Scotland, which follows a fault from Fort William to Inverness. (Map page 68.)

As well as the joints due to shrinking of the sediments, the stresses and strains due to movements in the earth's crust cause further fractures. Severe fractures may reach from the earth's surface deep down into the crust. These are called **faults.** As shown in the diagram on page 78 the strata on each side of a fault often move in such a way that particular sediments are no longer continuous. Faulting is of importance in coal-mining, for where the coal seams are broken and disturbed mining becomes difficult and expensive. When faulting is extensive and many faults occur close together, the strata may be dislocated to form **rift valleys.** The most famous example of a rift valley in Britain is the Central Lowlands of Scotland. (*See p. 78.*)

Dislocations of sediments in this way have been very widespread in Europe. Faults lie along the boundaries of many highland areas such as the Meseta in Spain, the Massif Central and the Vosges in France, and the Black Forest in Germany. In fact these highlands have been formed by movement following faulting and are called **block mountains** or **horsts.** Results of such movements are to be seen along the boundary between France and Germany. Here the Rhine Rift Valley lies between the block mountains of the Vosges to the west and the Black Forest to the east. The steep slopes of a rift valley are known as **fault scarps.**

14

You must remember that these scarps are formed very slowly.

Slight adjustments of this kind in the earth's crust cause **earthquakes,** which frequently lead to great damage and loss of life. Huge splits appear in the ground, down which men, animals, and even buildings have been known to disappear. As the ground heaves and shudders, whole cities are laid waste, as in Lisbon in 1755, San Francisco in 1906, and Hopei Province (China) in 1976, when 655 000 lives were lost. A girl caught by an earthquake while out cycling in Assam, India, wrote afterwards:

" When we had gone about 300 yards up the hill I suddenly heard a dreadful roar. I couldn't make out what it was. It sounded like a train, thunder, landslip all together, and it came nearer and nearer, and then the ground began to heave and shake and rock. I stayed on my bicycle for a second, and then fell off, and got up and tried to run, staggering about from side to side of the road. To my left I saw great clouds of dust, which I after-wards discovered to be houses falling and the earth falling from the sides of the hills. To my right I saw the embankment at the end of the lake torn asunder and the water rush out, the wooden bridge across the lake break in two, and the sides of the lake falling in; and at my feet the ground cracking and opening. I was wild with fear, and I didn't know which way to run."
(R. D. Oldham, *Memoirs of the Geological Survey of India*, vol. 29.)

Earthquake fissures in a road near Kiotu, Japan.

Here we see one of nature's most fearsome spectacles. When faults go deep enough into the earth's crust, cracks are provided through which molten magma can force its way up to the surface. When this happens a **volcano** is formed. This picture shows an eruption of the Mexican volcano Paricutin.

In many cases molten rock (**lava**), cinders and ashes are blasted up through the vent and soon build a cone, which in time may become several thousand metres high. The diagram explains this. Volcanoes go through active periods when eruptions occur, often with explosive violence, separated by dormant periods when lava ceases to flow. Damage to buildings and loss of life is sometimes heavy during eruptions, as at Mont Pelée (a volcano near St. Pierre in Martinique) in 1902, when an observer saw ". . . the volcano's side open with a rending, roaring explosion, and a vast cloud of dense black smoke shoot out at terrific speed. It was as if the other three sides of Pelée formed the mouth of a gigantic cannon from which shot a bolt of super-heated vapour, gas and white-hot fragments of molten rock.

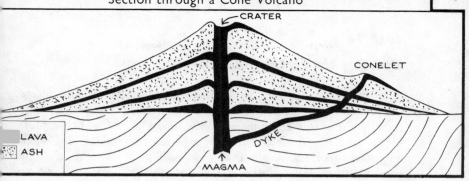

This tremendous blast rushed down the slopes of Pelée at 300 m.p.h., like a torrent of black fog. . . . The effect on St. Pierre was as if it had been hit by a hurricane composed of burning air heavily charged with red-hot sand and stones. The city was set ablaze from end to end almost instantaneously. . . . With the exception of two who miraculously escaped, St. Pierre's 28 000 inhabitants were wiped out almost before they knew what hit them." [1]

There are no active volcanoes in Britain today, but there are remnants of ancient ones, and volcanic rocks make up substantial parts of our islands. The famous Castle Rock in Edinburgh (see page 84) is the remnant of a former volcano. When volcanoes become extinct the molten lava in the vent goes solid; in time the crumbly ash which makes up the cone withers away under the attacks of the weather, and the harder " plug " sticks up on the landscape to form what is known in France as a **puy.**

Not all volcanoes are of the explosive variety. When the split in the earth's crust is wide and long, the lava oozes quietly out on to the earth's surface. Below sea-level, if the lava is thin and fluid, it will build up a broad, flat dome, like Mauna Loa and Kilauea in Hawaii: on dry land it will spread out to form a **lava plain.** The Snake River Plateau in the U.S.A., and part of the Deccan in India were formed in this way. Fragments of a former lava plain occur in the basalt flows of the north-east of Ireland and in the Hebrides. Notice in the sketch on page 227 the splits and joints in basalt, which are similar to those found in sedimentary rocks. You can see joints, too, in the photograph of granite on pages 9 and 210. These joints in igneous rocks are caused by the shrinking of the rock as it cools from the white-hot molten state.

[1] Frank W. Lane, *The Elements Rage* (Country Life).

CHAPTER 2
Moulding the Earth's Surface

" A GIANT MASS of rock which had threatened the Alpine village of Zambana near here for a month crashed down with a roar today—but spared the village. Villagers, who had spent sleepless nights watching the rock lit up by an army searchlight, fled when engineers gave the alarm that nine million cubic feet of stone were about to fall. Cattle stampeded, but the crash caused no casualties. As the thick dust gradually cleared, leaving the village and surrounding countryside shrouded in white powder, inhabitants found they might now be faced with a disastrous flood. The rock fell into two mountain rivers above Zambana which may sweep the huge mass of rubble down to bury the village.

" About half of Zambana's two thousand inhabitants have already left to take shelter in neighbouring villages, and hundreds of birds are lying dead under a blanket of dust." [1]

Landslides of this type, but not always on this spectacular scale, are a common occurrence in mountainous districts. They remind us that nothing in the world is really permanent. In the last few pages we have learned something of how mountains are built; our attention must now turn to the mountain *destroying* forces of **denudation** which are continually at work reducing even the highest mountains towards sea-level by **weathering,** i.e. the shattering of rocks by (*a*) alternating heat and cold; (*b*) frost; and (*c*) chemical decay due to acid rainwater; and by **erosion,** i.e. the gouging of surfaces by rock fragments as they are removed downhill.

When substances are heated they expand ; on cooling they contract. Under the alternating influence of heat from the sun, and cooling due to the sun's absence at night, rock surfaces are, in time, shattered. The fragments of rock are in turn broken down into smaller, and then still smaller, particles. This simple process provides one clue as to how mountains are gradually broken up and demolished by the forces of nature.

Water trickles into joints and surface cracks of the bare rock on mountain sides. At night it freezes, and as it freezes it expands with the same force that bursts our water-pipes in winter. Like thin wedges driven into the cracks, the ice splits the rock apart

[1] *The Guardian.*

until fragments break off and
tumble downhill. An excellent
example of this mountain-destroy-
ing force in action occurs on the
south side of Wastwater in Cumberland. Here, great **screes** of rock
fragments have collected on the
lower slopes of the mountain side.

This odd picture provides us
with another clue to the mountain-
destroying forces of nature. It
shows a carved stone head, standing
outside the Sheldonian Theatre, a
university building in Oxford. It

was placed there, together with other similar figures, in A.D. 1664.
No doubt at that time it was an excellent piece of sculpture—today
it is decayed and hardly recognisable. This is because it has partly
been dissolved away by rainwater, which is a weak acid.

Carbon dioxide gas from the air dissolves in rainwater to form
carbonic acid. Furthermore, humic acid from decaying plants
gets into the rainwater when it reaches the ground, and these
acids readily attack certain rocks, especially limestones (which
are carbonates). The result is that limestone areas are riddled
with holes and underground caverns where great masses of the
material have been dissolved away and carried off in solution to
the sea by the rivers.

As a result of weathering, loose rock fragments accumulate on
the ground surface. On barren mountain slopes or in deserts
these fragments are obvious. In most places, however, soils
develop on the broken rocks so that they become hidden from
view beneath vegetation and crops.

Unless they are covered with soil these fragments of rock seldom
remain stationary for long. Dislodged boulders tumble from cliff
faces; the wind blows away smaller particles in clouds of dust;
running water carries off soil, pebbles, and even boulders (during
floods) towards the sea; whilst moving masses of ice—in the shape
of glaciers and ice-sheets—are capable of shifting enormous
quantities of rubble. These various forces—gravity, wind, water
and ice—which move the materials produced by weathering are
called the **agents of transport.** Each of them has a special
and noticeable effect on the landscape.

These are the remains of what is believed to be the oldest church in England. They are near Perranporth, on the north coast of Cornwall, 1 km from the sea. The church, built by St. Piran, a Christian missionary from Ireland, in the 5th century, was at one time completely overwhelmed by the sand which at this point is continuously being blown inland from the beach. Notice the mounds (here partly overgrown) called **sand-dunes** surrounding the church on all sides. Wherever masses of sand occur, whether on the coasts or in deserts, dunes are built up by the wind.

Driven forward over the ground surface, sand particles pile up against obstacles like boulders in their path to form dunes which may be several hundred feet high. These dunes usually move in the direction of the wind. The grains of sand are blown up the gentle slope and fall down the steeper leeward slope. Hundreds of these hills of sand cover the Erg district of the Sahara, south of the Atlas Mountains. Because the sand-grains move quickest at the sides of the dunes they tend to become crescent-shaped, and are then known as **barkhans.**

Attempts are frequently made to arrest the movement of wind-blown sand, as it causes serious damage to agricultural land, and has been known to overwhelm whole villages. Marram grasses and conifers are planted to bind the sand-grains together. Noted attempts of this kind have halted the migration of sand inland at Culbin, on the Moray Firth, and in the Landes district of France.

The shattering of rock by alternating heat and cold (page 18) occurs most dramatically in the hot deserts. During the campaign in the Libyan desert in the Second World War many soldiers were astonished at the difference between the blistering heat of the day—when eggs could be fried on the bare rocks—and the bitter cold before dawn, when temperatures occasionally dropped below freezing point. The main reason for this large daily range of temperature is that at night the warmth of the desert floor is quickly radiated away into the atmosphere through the clear, dry,

desert air. These extreme temperature changes help break up
the rock (*How?*) into smaller and smaller fragments. At times,
temperatures fall low enough for dew to form, briefly coating rock
surfaces with a film of moisture. This water, together with that
from rare rainstorms (see below) rots the rock by chemical decay.
(See p. 19.) The broken and rotted pieces of rock finally form
the sand for which the deserts are famous. Over large tracts
of the desert, however, the sand has been blown away, leaving a
barren 'rock floor' littered with boulders and angular fragments.

Wind-blown sand acts as a natural sand-blast on all exposed
rock surfaces. Bit by bit, the softer rocks are eaten away. As
the sand-grains are more numerous and are travelling faster at
ground level, irregularities in the desert floor are gradually cut
away at their base. Rock pillars of extraordinary shapes develop,
in which differences in the hardness of strata show up remarkably.
Wind-erosion is responsible for many of the peculiarities of
desert scenery. In particular it helps to carve the steep slopes of
mountain remnants (*see diagram*). These rocky hummocks,
often tabular in shape, are called **inselbergs** ('island mountains').

Not all erosion in deserts is the result of abrasion by wind-
blown sand. Surprising as it may seem, *running water* also plays
its part in carving up parts of the desert floor. Few deserts are
entirely without rainfall. Thunderstorms with torrential down-
falls of rain occur at infrequent intervals. Naturally the rainfall
tends to be heavier and more frequent towards the fringe of the
desert lands. When the rains come the water sweeps in floods
down from the higher ground, for it runs quickly over the barren
rock surface since there is no soil or vegetation to soak it up.
Rock fragments carried along by the flood, and the force of the
water itself, dig deep trenches called **wadis.** Usually wadis are
quite dried up, but following a rainstorm they swiftly become
filled with a raging torrent.

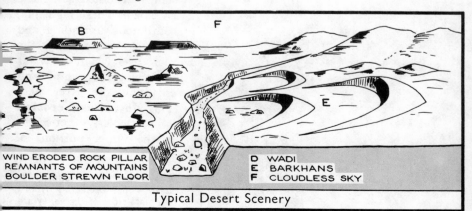

A WIND ERODED ROCK PILLAR	D WADI
B REMNANTS OF MOUNTAINS	E BARKHANS
C BOULDER STREWN FLOOR	F CLOUDLESS SKY

Typical Desert Scenery

This photograph was taken at Lynmouth (on the coast of north Devon) in 1952. During the night of the 14th August a portion of the village had been destroyed by flood. The River Lyn, normally a small stream, became a swirling torrent soon after exceptionally heavy rain had fallen on Exmoor. Great boulders and rock fragments were swept through a main street, destroying houses and drowning 28 people. These materials were brought down the Lyn valley, having been broken off farther upstream by weathering. The beds of streams in their steep mountain and hill courses are always littered with such fragments, varying in size from small pebbles to large boulders. Very large boulders are moved downstream only during times of flood, when the power of the river is enormously increased; but some material, especially the smaller pebbles, is constantly trundled along the river bed. These fragments, continuously grinding and scouring, wear away the rocks over which the water is flowing, deepening and enlarging the river bed. This **downward erosion** is particularly noticeable in the upper portions of

A V-shaped river valley. The steep slopes projecting from alternate sides of the valley are called **interlocking spurs.**

rivers, which become V-shaped as a result, as shown below, left. (The rock fragments are also worn smaller —take up a handful of pebbles from a stream bed and notice how rounded and smooth they are.)

Pot-hole on the bed of the River Spey, Inverness.

Stones falling into hollows in the rocky river bed are swirled around by the water to carve out **pot-holes** like those shown here. As more and more rock is cut away these pot-holes sometimes link together to form deep trenches. The tiny fragments chipped off add to the **load** of material which the river is bearing away to the sea.

The swift upper course of a river is often broken by numerous **waterfalls, rapids** and **plunge-pools**—the latter being formed at the foot of the waterfalls as pot-holes are enlarged by the swirling boulders they contain. The waterfalls and rapids mark sections of the river bed more resistant to erosion.

Farther downstream the **gradient** (the slope of the river bed towards the sea) becomes less steep and irregular, and the flow therefore becomes more uniform in speed. Increased in size by the **tributaries** which have joined it, the stream is less affected by sudden, dramatic changes in volume and its load consists mainly of small particles of mud and rock. It is now eroding its banks sideways as much as it is cutting its bed downwards; thus the V-shape of its cross-section becomes much broader.

Wherever the current slackens, the river now tends to drop some of the load it has brought down from the mountains. Patches of gravel, sand and silt accumulate to build up **alluvial flats.** These are often of great value as they provide rich meadows for farming in otherwise hilly districts.

23

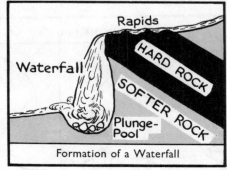

Rapids

HARD ROCK

Waterfall

SOFTER ROCK

Plunge-Pool

Formation of a Waterfall

Flood plain of R. Cuckmere, Sussex.

(*1*) Explain, from the diagrams below, why it is unwise to stand very close to the edge of a river bank on the *outside* of a bend. Such bends (**meanders**) develop for reasons that are not yet fully understood. They may be found at any point on a river's course, but are particularly common where it flows across alluvial flats. The flow is swiftest on the *outside* edge of these bends and loops, so that the outside bank is constantly undercut by **lateral erosion** (*see below* (*2*)). Where the current runs more sluggishly on the *inside* of the bend **deposition** occurs and the bed is slowly filled in as shown.

The meander thus grows in size until the river has bent so far back on itself that only a narrow neck of land separates two parts of its course. During a flood this narrow strip is frequently broken across so that the river's course is straightened, and the former loop no longer takes the main flow of water. (*2*) Make a sketch of the photo, and show by an arrow where this is likely to occur. In time the **cut-off,** or **ox-bow lake**, silts up and becomes overgrown.

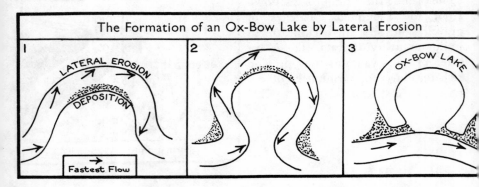

The Formation of an Ox-Bow Lake by Lateral Erosion

Farmers living on the neck of land of large loops on the River Mississippi dug ditches across the loop, during floods, to divert the course of the river. By bringing the river through their plantations they made it easier to market their produce by boat and greatly increased the value of their land. The following description reminds us of the power of a river the size of the Mississippi:

" When the water begins to flow through one of those ditches . . . it is time for the people thereabouts to move. The water cleaves the banks away like a knife. By the time the ditch has become twelve or fifteen feet wide, the calamity is as good as accomplished, for no power on earth can stop it now. When the width has reached a hundred yards, the banks begin to peel off in slices half an acre wide. The current flowing around the bend travelled formerly only five miles an hour; now it is tremendously increased by the shortening of the distance." [2]

The bed and banks of very large rivers, in their lower course, are gradually raised as alluvium is dropped by receding flood waters. Repeated every flood season for many years, this process eventually leads to the peculiar situation where the river flows at a level *above* that of the **flood plain** it crosses. This is a very dangerous state of affairs, for the **levées,** as the natural embankments on each side of the river are called, are liable to give way and lead to serious flooding.

At the river's mouth the flow may in some cases be strong enough to sweep most of the silt well out to sea. Where the flow is weak *in relation to the load carried* the silt is checked or even washed back into the river's mouth by the twice-daily incoming tides, and a **delta** is formed. The river then reaches the sea through a maze of small channels, separated from one another by flat mud islands. As more mud accumulates, the delta grows and spreads seawards. Deltas hinder shipping, but the mud flats are often very fertile.

There are deltas in some British lakes, e.g. Lake Bala (Wales) and Lake Derwentwater (Cumberland), where streams entering the lakes deposit their load. (See also page 105.)

Seaward End of Mississippi Delta

² Mark Twain, *Life on the Mississippi.*

Deep below Ingleborough (a high part of the Pennine Hills in Yorkshire) is an underground cavern large enough to contain the great York Minster cathedral. This enormous chamber, the largest of hundreds of similar cavities in the area, has been formed through millions of years by rainwater dissolving away the limestone rock of which the base of Ingleborough is mostly composed.

Solution of the rock always plays a very important part in the erosion of limestone uplands. All limestones were laid down originally as sediments under water, most of them by the accumulation of millions of shells and skeletons of sea-creatures. The shells are made of calcium carbonate, and on page 19 we learned that this material is readily dissolved by the weak acids in rainwater. The joints in limestone provide easy paths along which rainwater can attack the rock. Underground tunnels are formed as it is dissolved away. As a result there are few streams at the surface in limestone districts, for the rainwater soon trickles down crevices, and streams disappear down '**swallow holes**'. At times, so much limestone is dissolved that the roofs of connected caverns collapse to form a **gorge**. (See photo, page 219.) Vegetation is very scanty on the thin, waterless, limestone uplands.

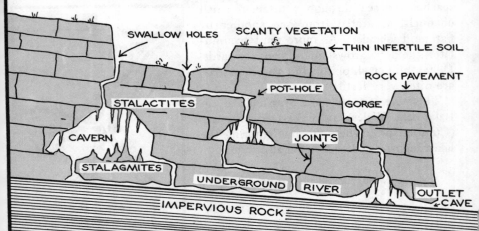

Features of Limestone Country

FEW SURFACE STREAMS : INTERMITTENT DRAINAGE

SWALLOW HOLES
SCANTY VEGETATION
←THIN INFERTILE SOIL
ROCK PAVEMENT
STALACTITES
POT-HOLE
GORGE
CAVERN
JOINTS
STALAGMITES
UNDERGROUND
RIVER
OUTLET
CAVE
IMPERVIOUS ROCK

There are often no trees, but only a few coarse grasses. In many places there are no plants at all, and bare **'rock pavements'** form the ground surface. (See photo, page 110.) This type of limestone scenery is called **karst**.

Below ground the water containing dissolved limestone sometimes trickles so slowly that it evaporates, leaving the solid carbonate behind. This happens, for example, when water drips slowly from the roof of a cavern. An 'icicle' of limestone—called a **stalactite**—grows down from the roof, whilst a similar pillar—a **stalagmite**—grows upward from drips that reach the floor of the cave. It takes thousands of years to form pillars such as that shown below.

The line of hills in this photograph is the Cotswolds in Gloucestershire. (*3*) Find them in your atlas. These oolitic limestone hills present a steep or **scarp** slope towards the north-west: the much gentler descent towards the south-east is called the **dip** slope. Running parallel to the Cotswolds some 65 km farther southeast are the chalk hills of the Chilterns. These hills also have distinct scarp and dip slopes. Between the Cotswolds and the Chilterns lies a broad plain occupied by the Vales of Oxford and Aylesbury, where clay is the main rock. (See page 173.)

At one time these chalks, clays and limestones were horizontal sediments on the bed of a shallow sea. After being uplifted to form dry land they were tilted towards the south-east and then carved by river erosion into their present shape. (*4*) Look carefully at the diagram and see if you can write down the following paragraph correctly:—

As limestone and chalk are both $\frac{\text{PERMEABLE}}{\text{IMPERMEABLE}}$, rainwater rapidly $\frac{\text{RUNS OFF}}{\text{SINKS INTO}}$ them, and $\frac{\text{LITTLE}}{\text{MUCH}}$ erosion by running water at the ground surface can take place. There are $\frac{\text{FEW}}{\text{MANY}}$ streams on the clay, however, for water $\frac{\text{CANNOT}}{\text{CAN}}$ sink through it. As the $\frac{\text{CLAY}}{\text{PERMEABLE ROCK}}$ is worn down much more $\frac{\text{SLOWLY}}{\text{QUICKLY}}$ than the $\frac{\text{CLAY}}{\text{PERMEABLE ROCK}}$, it stands out on the landscape as bold lines of hills called **escarpments.***

This map shows the many escarpments in south-east Britain. (*5*) From it name two escarpments consisting of (*a*) chalk; (*b*) limestone and other rocks.

* Also known as **cuestas.**

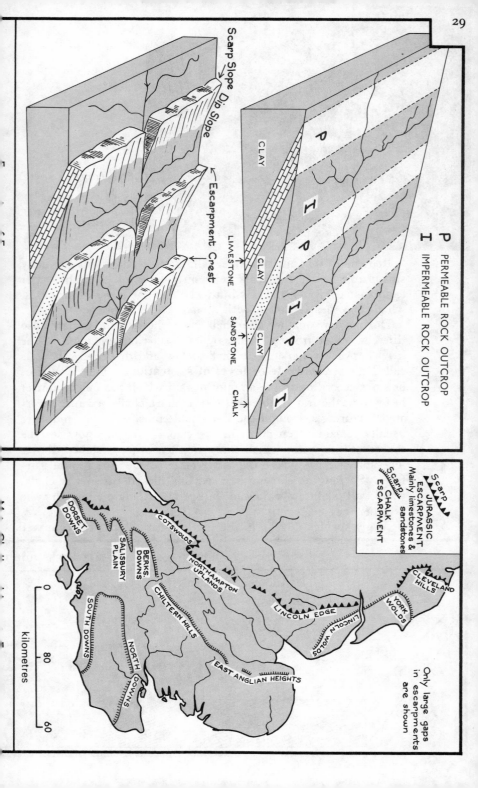

Scarp Slope

Dip Slope

Escarpment Crest

LIMESTONE

SANDSTONE

CHALK

CLAY

CLAY

CLAY

P   PERMEABLE ROCK OUTCROP

I   IMPERMEABLE ROCK OUTCROP

P   I   P   I   P   I

Scarp
JURASSIC
ESCARPMENT
Mainly limestones &
sandstones

Scarp
CHALK
ESCARPMENT

DORSET
DOWNS

SALISBURY
PLAIN

BERKS.
DOWNS

COTSWOLDS

NORTHAMPTON
UPLANDS

LINCOLN EDGE

LINCOLN WOLDS

CLEVELAND
HILLS

YORK
WOLDS

SOUTH
DOWNS

NORTH
DOWNS

CHILTERN HILLS

EAST ANGLIAN HEIGHTS

kilometres

0

80

60

Only large gaps
in escarpments
are shown

Even during the driest summers this spring at the foot of the South Downs has never been known to stop flowing. From where does the water come?

Remember that any rainwater falling on to chalk will gradually sink through. If the chalk rests on a layer of clay, the water cannot sink through the clay but will fill up the pores in the chalk, which becomes saturated with moisture like a sponge.

During dry weather the chalk near the ground surface also dries, as water from previous rainfall seeps underground. The level between water-logged chalk below and dry chalk above is called the **water table** or **level of saturation.** This level rises or sinks according to the amount of rain which has fallen. After heavy rain the water table will rise as more chalk becomes saturated. During a drought the water table falls, but some moisture usually remains deep down in the chalk. We can now see the source of the water which flows in the spring sketched above. The spring starts where the water table reaches the ground surface. So long as some water remains locked up in the chalk, water will continue to flow in the spring, no matter how dry the weather may be.

Many springs flow from the foot of scarp slopes or from well

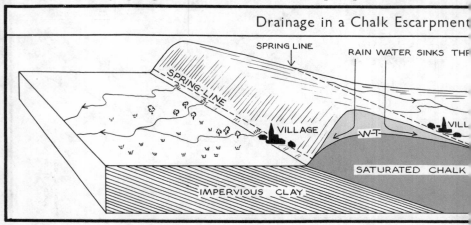

Drainage in a Chalk Escarpment

SPRING LINE   RAIN WATER SINKS THR

SPRING-LINE

VILLAGE   W-T   VILL

SATURATED CHALK

IMPERVIOUS CLAY

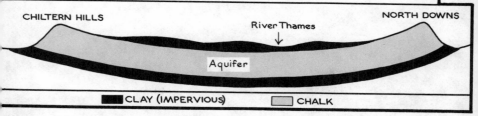

down the dip slopes near the junction between the porous and non-porous rocks. The places where the springs appear are said to lie on the **spring-line**. Spring-lines are frequently marked by villages—built there because drinking water was scarce farther up (above the water table) whilst the lowlands were too wet and marshy.

It sometimes happens that rainwater sinking into porous rock can find no outlet, and becomes trapped. This is the case in the chalk which lies beneath London.

The diagram above shows how the chalk of the Chiltern Hills dips below London and reappears as the North Downs. Above and below the chalk is clay, through which water cannot flow. Rain, falling on the chalk hills north and south of London, seeps down and saturates the thick layer of chalk beneath the capital. By boring through the surface layer of clay into the chalk beneath, a constant supply of pure water may be obtained. This is known as an **artesian** water supply. So much water has been pumped to meet the needs of London's inhabitants that the water table there has fallen. A rock that holds water in this way is called an **aquifer.**

In south-east England chalk is the most valuable aquifer, but elsewhere sandstones and limestones are locally important. In Nottinghamshire, for example, much water for the industrial towns is pumped from the Bunter Sandstone.

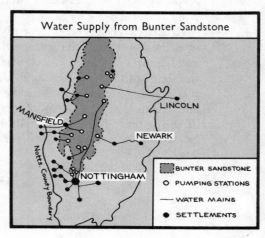

This **ice-cap**, covering most of Greenland, has been built up
by thousands of years of heavy and frequent snowfalls in the
central mountains of that island.   Snowdrifts hundreds of metres
thick are heavy enough to compress the lower layers of snow into
ice.   This ice then slides downhill from the mountains until it
melts in the warmer temperatures at or near sea-level.   On its
journey to the sea the ice pushes through former river valleys, and
the 'ice-rivers' which result are called **glaciers**.   In the photograph a glacier reaches the sea and large masses of ice have broken
away to form **ice-bergs**.

20 000 years ago (very recently, in geological time) the whole
of Britain north of a line joining the estuaries of the Rivers Severn
and Thames lay beneath a similar ice-sheet.   We know this from
the shape of the land surface in this part of Britain, especially the
mountainous districts of the Lake District, the Scottish Highlands
and North Wales.   The photograph on the next page shows part
of the latter.   Notice especially the U-shape of the valley—the
lower slopes are very steep, and the floor broad and nearly
level.

Llyn Bach Pass, Wales.

This is the shape typical of a valley which once contained a glacier. Compare it with the upper or mountain course of the river valley shown on page 22. As a glacier grinds its way down such a valley it exerts tremendous force against the walls and floor. Rock fragments, varying in size from giant boulders to small pebbles, are broken off and carried along in front of and beneath the advancing ice. Using these fragments as 'tools' for further gouging and scraping, the glacier gradually changes the valley's shape from a V to a U. Thus a glaciated valley is said to be **over-deepened.**

Many waterfalls plunge over the steep walls of a glaciated valley. These mark the places where small tributary glaciers once joined the surface of the main flow of ice. When the ice melted, the tributary streams were left 'stranded' high above the main valley floor, and now occupy **hanging valleys.**

At the head of many glaciated valleys is found an arm-chair-shaped hollow called a **cwm** (pronounced *coom*) in Wales, a **corrie** in Scotland, and a **cirque** in France. Such hollows are believed to be caused by long-continued frost erosion beneath the **snow-collecting fields** high up on the mountain slopes. After the ice melts, cwms often contain small lakes.

Section through
a Cwm

LIP OF CWM   CWM LAKE

FROST-SHATTERED REAR WALL

A cwm and cwm lake may be seen in this photograph of Mount Snowdon (1095 m) in North Wales. Notice, too, the narrow ridges, called **arêtes**: formed when two adjoining cwms bite into the mountain. The best climbing routes to a mountain summit frequently follow arêtes.

Compared with flowing water, glaciers move extremely slowly. In a day some move less than 1 metre, but others might travel as much as 20 metres. In 1956 the body of a German climber was found at the foot of the Grindelwald glacier in Switzerland. He had last been seen setting out to explore the glacier in *1887*. No doubt he had fallen into a deep split, or **crevasse**, in the glacier, been frozen into the ice, and gradually carried downstream.

Sometimes markings and grooves are found on rock surfaces over which fragments of stone have been scraped by the moving ice. The position of such **striations** helps to show the direction in which the ice travelled.

The dark lines on the glacier shown opposite are made by piles

Aletsch Glacier, Switzerland.

of rock fragments which, dislodged by frost, have tumbled on to the ice from the surrounding mountain slopes. Such mounds of rock are called **moraines.** Those at the glacier's sides are **lateral** moraines. Running down the middle are **medial** moraines, formed by the lateral moraines off separate glaciers which merge together. With the boulders and rocks dislodged from the valley floor and carried along within the ice, those of the lateral and medial moraines are dumped at the front of the glacier to form a **terminal** moraine. Similar moraines are built at the fringe of an ice-sheet.

If the climate of an ice-covered region becomes warmer, ice-sheets and glaciers gradually melt away. The lowland areas which formerly lay beneath the ice are left covered by masses of rocks and clay, called **boulder clay** or **ground moraine.** The clay consists of **rock flour,** powdered rock scraped off highland slopes and ground very fine. It often forms very fertile soil, and the boulder clay areas are among the best farming land in Britain. Notice overleaf and on page 105 the striking contrast between the glaciated mountains and the more fertile lowland where boulder clay has been dumped. (See also page 162.)

A glaciated valley in Perthshire. Note the contrast between the barren, scoured uplands and the more fertile valley floor, where boulder clay has been deposited.

Beneath a glacier or ice-sheet much rock is pulverised into very fine particles. In time this finely ground material collects at the terminal moraine. When dry, it is readily blown away by the wind. For thousands of years during and after the ice-age large quantities of this fine dust, known as **loess,** were blown southwards from the fringe of the Scandinavian and British ice-sheets. Where loess came to rest it formed a mantle of varying thickness which has now formed extremely fertile soils.

Not all glacial deposits are fertile. In some parts of Britain and the North European Plain large quantities of pebbles and sand were spread over the ground surface by the swirling waters from the melting ice-cap. Today, as in the Breckland in Norfolk, and the Luneburg district in North Germany, these sands and gravels form poor, infertile land on which few crops will grow. They remain as open heathland, or are covered with coniferous trees which like sandy conditions. (See photo, page 163.) The gravels are useful, however, as road-building materials and for the making of concrete.

This diagram shows the main features of a glaciated landscape. (6) Make a large copy of it, adding the correct name (from the following list) of each lettered feature.

terminal moraine	sand and gravel	boulder clay
truncated spur	cwm	waterfall
hanging valley	lateral moraine	arête
rounded skyline	broad, overdeepened valley floor	

Add arrows to show the direction in which the ice moved.

Houses undermined by cliff falls at Robin Hood's Bay, Yorkshire.

(See map, p. 143.)

On the coast a constant battle is being waged between the waves and the dry land. At times the sea triumphs, as along the Yorkshire coast where the top photograph was taken. Elsewhere the land is gaining ground from the sea. What are the forces involved in this struggle?

Sea-waves have enormous power. When they break, thousands of tonnes of water are hurled against the cliffs which, in time crumble under the strain. In cold weather, too, frost action helps

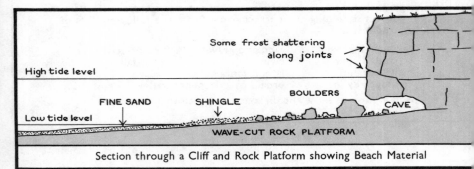

Section through a Cliff and Rock Platform showing Beach Material

to shatter the cliff face. The dislodged rock fragments are then used by the waves as 'tools' for further erosion. Washed to and fro, and thrown about during storms, they grind and scour away at the foot of the cliffs. Any weaknesses in the rocks, such as joints, are opened up and caves are formed (photo, page 142). These undermine the cliffs and eventually lead to more rock falls. Under prolonged erosion of this kind the cliffs gradually retreat, leaving a bare **rock platform,** covered in places with a mantle of rock waste. Compare the diagram with the photographs on pages 12 and 196.

The rock fragments are themselves gradually broken into smaller pieces and become rounded by being ground against one another and against the cliff walls. At the foot of many cliffs it is possible to find rock fragments varying in size from huge boulders to tiny grains of sand. The latter often consist of finely ground quartz (silica)—the very hard, resistant, igneous rock mentioned on page 8.

The smaller pebbles and sand particles are washed out to sea, or along the coast. Movement *along* the coast, known as **longshore drift,** is particularly important. It gradually fills up hollows and indentations in the coastline, building beaches and sand bars. The photograph below shows Orford Ness, a sand and shingle **spit** built by longshore drift along the Suffolk coast, south of Aldeburgh.

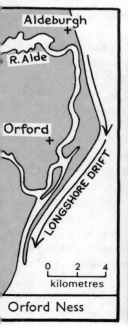

Orford Ness

Spits similar to Orford Ness are found at many places around the coasts of the British Isles. This photograph shows Chesil Beach, a shingle bank which links the former island of Portland with the Dorset coast. During severe storms the railway and road running along the spit are blocked by floods, for seawater comes over the beach and through the pebbles, temporarily making Portland Isle a true island again.

In Mount's Bay, on the south Cornish coast, a shingle bar built by longshore drift completely blocks the estuary of the River Cober. Helston was once a small sea-port: (7) How far is it now from navigable water?

Along some parts of the British coast there are remnants of ancient beaches, several feet above the level reached by even the highest tides today. These **raised beaches** are very plentiful along the shores of southern England. On some coasts, too, it is possible during low tides to see tree-stumps on the sea-bed—the remnants of former

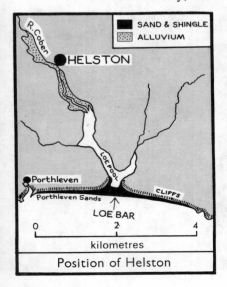

Position of Helston

40

forests. The tree remains shown in this photograph are on the beach near Portscatho in south Cornwall.

From evidence such as this we know that the level of the sea is by no means constant—apart altogether from its twice-daily rise and fall due to the tides. Changes in sea-level can result from a gradual sinking or rising of the land, or from changes in the amount of water in the ocean. The melting of the Greenland and Antarctic ice-caps, for example, would raise by 60 metres the present level of the ocean waters and put most of London on the sea-bed. Such changes, of course, take place extremely slowly, but continued for a long time their effects are important.

When the sea-level rises, the coastline becomes very uneven. Arms of the sea reach far inland, drowning former river valleys, and forming **rias.** This has happened along the coasts of Devon and Cornwall, Pembroke and south-west Ireland. Rias make fine harbours, and many ports are found on them, e.g. Devonport, Milford Haven, Cork and Falmouth (photo, page 212).

When a rising sea-level invades a glaciated highland region the drowned *glaciated* valleys are known as **fiords.** There are many fiords on the west coast of Scotland.

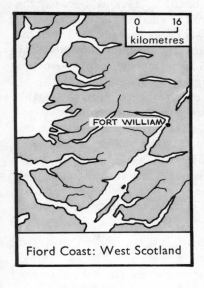

Fiord Coast: West Scotland

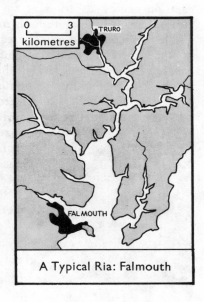

A Typical Ria: Falmouth

CHAPTER 3

The Weather

A STRONG GALE wrecks ships, damages houses, blows down trees and hinders transport; heavy rains wash away soil and cause floods; droughts destroy crops and cause famines—in these, and countless other less dramatic ways, the weather has an impact on human affairs. No people are more aware of the weather than we in Britain. " Nice day "—" Cold wind "—" Fine evening " —" Brighter now . . ." we greet one another with these and endless similar remarks. The weather matters so much to us that scientists called **meteorologists** make a constant study of it. They make accurate observations of the **atmosphere**—the invisible layer of mixed gases, several hundred kilometres thick, which completely envelops the solid earth.

One of the atmospheric gases is oxygen, which we all breathe and depend upon for life. Normally we can easily obtain all the oxygen our bodies want, but on high mountains the slightest exertion causes us to pant for breath. This is because the atmosphere at high altitudes is more rarefied: its density is lower, and less oxygen—and other gases—is present in any given volume of air. In fact about one-half of the whole atmosphere (by weight) lies within 6 km of the earth's crust, and above this height it thins rapidly. The diagram opposite shows this.

The weight of the atmosphere exerts a force on to the earth's surface called **atmospheric pressure.** At sea-level it is a little more than 1 kg/cm², but it decreases steadily with altitude. This pressure is much less at the summit than at the foot of a high mountain, for there is much less air pressing down.

Dougal Hasten on Mt. Everest in full climbing kit. Note the oxygen mask.

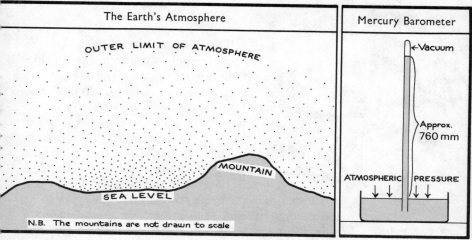

OUTER LIMIT OF ATMOSPHERE

MOUNTAIN

SEA LEVEL

N.B. The mountains are not drawn to scale

←Vacuum

Approx. 760 mm

ATMOSPHERIC PRESSURE

Atmospheric pressure is usually measured with a mercury barometer. The weight of air above the surface of the mercury in the trough forces a certain amount of mercury into the vacuum tube. If the atmospheric pressure decreases, the mercury column falls; if the pressure increases, the column rises. At sea-level it normally varies between 735 and 785 mm, but the meteorologist records pressure in **millibars,** with 1000 millibars equalling 750 mm on the mercury barometer.

At any one place on the earth's crust the atmospheric pressure is constantly altering, and it is because of this that winds blow. Air always moves from where the atmospheric pressure is high to where it is lower. The speed and force of the wind depend upon (a) the difference in pressure and (b) the distance between the centres of high and low pressure. When pressures are very different over short distances, gale-force winds develop: small differences in pressure result in gentle breezes.

Weather observers throughout the British Isles, and on ships around our coasts and in the North Atlantic, send daily reports to the Air Ministry Meteorological Office in London. Their reports include information on atmospheric pressure, temperature, winds, visibility, the dampness of local air, cloud types and rainfall.

Hundreds of pressure recordings are plotted on a large map, and lines are drawn *linking all places with the same atmospheric pressure at that time* (after allowance has been made for altitude). Such lines are **isobars** (see maps, pages 59 and 61) and the pattern they form is of great value in helping to forecast the direction and strength of winds, as well as other weather conditions.

The Eskimos wear thick furs to keep themselves warm, but people like the Australian aborigines need no such protective clothing. Why is it so bitterly cold near the Poles, whilst lands near the Equator swelter in the heat? Why, for that matter, is it distinctly warmer during the summer months along the south coast of Britain than in northern Scotland? The main reason lies in the angle at which the sun's rays strike the earth's surface. Look carefully at the diagram below. Within the tropics the sun is always high in the sky. The nearer one gets to the Poles, the shallower is the angle at which the rays reach the earth.

The two rays *A* and *B* are of equal width, and have the same heating power. Ray *A*, at the Equator, concentrates this power on to a relatively small portion of ground; ray *B* spreads the same amount of heat over a very much larger area near the North Pole. The ground surface near the Equator therefore gets much hotter than that near the Pole. Furthermore, some heat from the sun's rays is absorbed by dust and moisture particles floating in the air. (*I*) Can you explain, from the diagram, why the rays near the Poles lose most heat in this way?

(*2*) You can carry out a simple experiment along these lines. Arrange a cardboard shield over an electric torch so that the rays of light are sent directly forward. Now shine it on to a large ball, first directly on to its 'equator', and then—along a parallel path—on to one of its 'poles'. Notice the difference in the brightness and size of the patches illuminated. What does this show?

It is common knowledge that as one ascends a mountain the air gets colder. In fact the temperature drops about 1° C. for every 150 metres one climbs. Tourists can find it snowing heavily at the summit of mountains like Snowdon, in North Wales, having climbed from warm and pleasant conditions below. At first sight this seems strange: the explanation is this:—

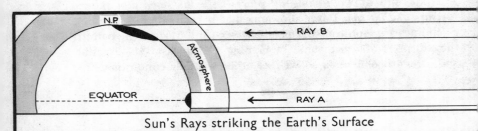

Sun's Rays striking the Earth's Surface

1. Heat rays from the sun pass right through the atmosphere, scarcely warming it;

2. The ground surface *is* warmed by the sun's rays;

3. The warm ground then gives off heat which *does* warm the air;

4. The air is warmed mostly where it comes into contact with the warm ground. *Thus the air is heated from below;*

5. At high altitudes the air is less dense (look again at the diagram on page 43), and so there is less of it to be warmed. Also there are fewer dust particles high up to absorb the heat.

Partly because of the low temperatures, no trees will grow on mountain slopes above a certain height, called the **tree-line.** Above this line only mosses, lichens and a few hardy alpine plants can survive. Higher still, the mountain surface is barren rock, often snow-clad, with no trace of vegetation. Make a sketch of the photograph of part of the Scottish Highlands on page 36, marking in the tree-line. Large parts of the Highlands, and smaller portions of high land elsewhere in Britain, are too cold and inhospitable for people to live there permanently.

Air temperatures are recorded *in the shade* by thermometers kept in a special cabinet called a **Stevenson screen.** This stands on legs 1·23 m above the ground surface and, as the photograph shows, has shutters which allow air to circulate freely around the thermometers, but shelter them from the direct rays of the sun. (3) Why do you think Stevenson screens are painted white?

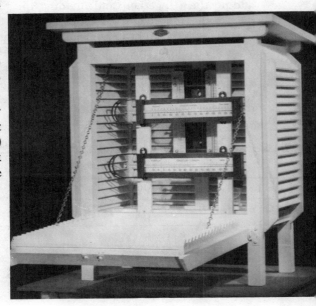

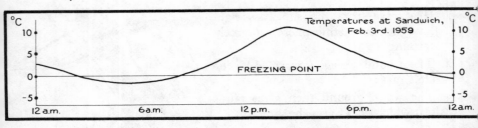

The temperature of the air at any one place is constantly varying, as the chart above shows. Generally speaking, the hottest part of the day is the early afternoon, and the coldest time is just before dawn. The highest and lowest temperatures during a day are recorded and the mean, or average, of these two readings gives the **mean temperature** for that day. For example:—

Maximum temperature for day............20° C.
Minimum temperature for day............7·8° C.

Therefore mean temperature for day $= \dfrac{20 + 7\cdot8}{2} = 13\cdot9°$ C.

(4) Work out the mean temperature for the day for which the graph was drawn.

By finding the mean daily temperatures for a whole month, adding them together, and dividing the total by the number of days in the month, a mean temperature for that month is obtained. But the month might be unusually warm or cool in one particular year. To offset this difficulty an average is taken of the mean temperatures for that month in at least 35 years, if these records are available. The final figure is then called the **mean monthly temperature.** (5) From the figures given below, work out the

Maximum and Minimum Daily Temperatures at Cambridge: January, 1958																														
Day	1	2	3	4	5	6	7	8	9	10	11	12	13	14	15	16	17	18	19	20	21	22	23	24	25	26	27	28	29	30	31
Max. (° C.)	8	4	4	9	8	13	7	10	8	8	7	4	4	5	7	8	9	9	4	3	1	1	2	2	6	10	13	9	9	8	4
Min. (° C.)	4	1	-1	-1	1	2	2	1	2	2	4	-2	1	1	-2	4	6	-1	2	-3	-5	-6	-6	-8	-6	-2	5	7	7	-1	-1

Monthly Mean Temperature at Cambridge in January, 1924-1958									
	°C.		°C.		°C.		°C.		°C.
1924	4·4	1931	3·4	1938	5·6	1945	0·4	1952	3·1
1925	4·9	1932	6·1	1939	4·7	1946	3·2	1953	3·1
1926	3·7	1933	2·1	1940	−1·5	1947	1·6	1954	2·7
1927	4·2	1934	3·8	1941	0·6	1948	3·6	1955	2·2
1928	4·8	1935	4·2	1942	0·4	1949	3·2	1956	3·4
1929	1·4	1936	3·8	1943	4·7	1950	4·4	1957	5·5
1930	6·1	1937	5·0	1944	6·1	1951	4·1	1958	?

mean temperature for January, 1958, at Cambridge.

(6) Using the additional information above, work out the mean January temperature at Cambridge.

In countries like Britain, where temperature alters considerably throughout the year, the mean monthly temperatures for January and July provide useful guides to general conditions of heat and cold. Care must be taken, however, not to rely too much on mean temperature figures: unless you have your wits about you they can be misleading. (7) For example, the chart shows the mean daily temperature for Sandwich on February 3rd 1959 was *above* freezing, but for how many hours during that day was the temperature below freezing point? Accurate information about the possibility of frost is particularly important to farmers, for many crops (fruits, especially) are damaged when it freezes.

As a *rough* guide to help you imagine conditions when particular air temperatures are mentioned, study the following notes:—

	°C.
Very Hot	30+
Hot	20 to 30
Warm	10 to 20
Cool (or Mild)	0 to 10
Cold	−10 to 0
Very Cold	Below −10

In temperate latitudes, as in Britain, the sun is never overhead. Even in summer the sun's rays *slant* in towards level ground. Look carefully at Diagram *A* overleaf.

(8) Answer the following questions:—

    1. Where are the sun's rays striking the ground vertically?
    2. Which slope never gets any direct sunlight?
    3. Which is receiving the more concentrated heat—the level
       ground, or the slope facing the sun? (Compare this
       diagram with that on page 44.)

(9) Copy the diagram and, assuming that it represents a valley in Britain running east-west, add the following labels in appropriate places:—

No direct sunlight	North-facing slope
South-facing slope	Strongest heating here

Rays striking ground vertically

    The direction in which a piece of ground slopes is thus important in deciding how much heat it receives from the sun. In Britain, farmers favour sheltered, south-facing slopes for the planting of crops requiring abundant sunshine. In Cornwall such slopes are planted with early spring vegetables and flowers (see picture, page 210). Inland from Dundee, in the Carse (garden) of Gowrie, they are lined with fruit farms (see page 77).

    Fruit farmers must also bear in mind that occasionally the air temperature *increases* with altitude: a state of affairs known as **temperature inversion.** (*10*) Look at Diagram B and explain why valley floors often get more frost than the hill slopes above. (*11*) If you were a fruit farmer in the valley shown in A, where would you plant trees: at A, or B, or C? (*12*) *Why?* Temperature inversion occurs only in calm weather. (*13*) *Why?*

    Temperature inversion was responsible for the ' smogs ' which once had such deadly effect on health in London, the Midlands and other industrial centres. Hot air, laden with smoke and poisonous fumes from multitudes of chimneys, at first rose swiftly through the cold, heavy air near the ground surface. It was rapidly chilled, however, and when it reached the ' ceiling ' of warmer, less dense air above it could not rise higher. It then spread out and formed a blanket of smoke and grime over the landscape. Provided the weather remained calm, the ' smog ' prevented the sun's rays from reaching and heating the ground. Thus the polluted atmosphere could not be dispersed by convection currents, and sometimes it lasted for several days, as in London ' pea-soupers '. (*14*) What steps have the Government taken to try to prevent ' smogs '? (*15*) Will nuclear-electrical power help to prevent them?

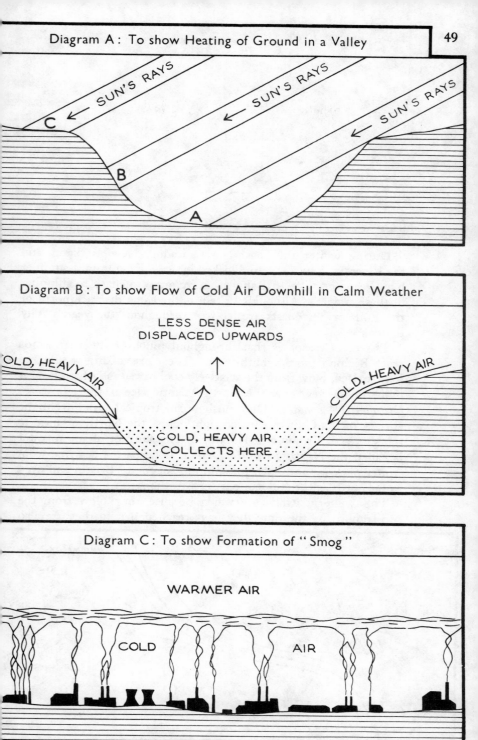

Diagram A : To show Heating of Ground in a Valley

SUN'S RAYS

SUN'S RAYS

SUN'S RAYS

C

B

A

Diagram B : To show Flow of Cold Air Downhill in Calm Weather

LESS DENSE AIR
DISPLACED UPWARDS

COLD, HEAVY AIR

COLD, HEAVY AIR

COLD, HEAVY AIR
COLLECTS HERE

Diagram C: To show Formation of " Smog "

WARMER AIR

COLD   AIR

During winter the coasts of Labrador, Newfoundland and the River St. Lawrence are blocked by ice. Even in mid-summer huge icebergs, like that shown above, are a menace to shipping in these waters. Although Britain is the same distance north of the Equator her coasts are ice-free throughout the year. Why is this?

The main reason is that throughout much of the year winds reach Britain from the south-west. These **prevailing winds,** the **Westerlies,** blow from the relatively cool **air mass** which lies for much of the year over the North Atlantic Ocean. ((*16*) *In which latitudes? (See map.)*)) They push warm tropical water from the surface of the Gulf of Mexico right across the North Atlantic to the shores of Britain and Norway. This current of warm water is called the **North Atlantic Drift.** It keeps the sea free from winter pack-ice and open for shipping all round Britain and even as far north as the Russian port of Murmansk.

The westerly winds are warm because they blow from the sub-tropics; by the time they have reached Britain they are also

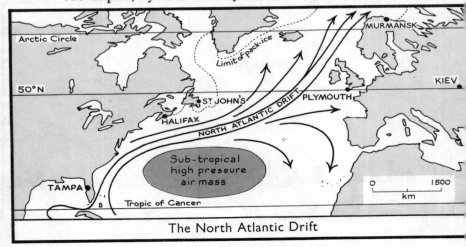

The North Atlantic Drift

very moist, having absorbed much water from the surface of the North Atlantic. This warm, moist westerly **airstream** is responsible for keeping winter air temperatures mild in and around Britain, compared with those on the other side of the Atlantic Ocean.

In winter the land cools much more quickly than the sea, and so the air inland tends to become much colder than that over the oceans in the same latitude. In summer, when the land heats more rapidly than the sea, air temperatures are usually higher inland. Overall differences in air temperatures in the course of a year—the **annual temperature range**—are thus greatest at places far inland.

In Britain the sea has a moderating influence on air temperatures throughout the year. Warm westerly winds off the Atlantic keep winter temperatures higher than would otherwise be the case. In summer the on-shore winds have a *cooling* effect, preventing temperatures from rising abruptly as they do, for example, in Central Europe. Having no extremes, either of heat or cold, British climate is said to be **equable.**

(*17*) Sketch a *large* copy of the map from the opposite page. Then make temperature charts for Tampa, Halifax and Plymouth, and stick them on your map in their correct places. (*18*) Describe and account for the differences in the January temperatures in these places.

(*19*) Make a temperature graph for Kiev and add it to your map. State:—

(*a*) the January temperatures in Plymouth and Kiev; and
(*b*) the annual temperature range in each of these places, and explain the difference.

On the maps overleaf the mean monthly temperatures at various places in Britain are compared during January and July. The lines on the maps are **isotherms,** i.e. *lines linking places which have the same mean temperatures* after allowance has been made

Mean Monthly Temperatures

		Jan.	Feb.	Mar.	Apr.	May	June	July	Aug.	Sep.	Oct.	Nov.	Dec.
Tampa	°C.	15·8	16·1	18·9	21·6	24·1	27·2	27·5	27·5	26·7	23·3	18·9	16·1
Halifax	°C.	−2·5	−4·4	−0·8	4·4	9·7	14·7	17·8	18·3	14·4	9·4	3·3	−2·8
Plymouth	°C.	6·4	6·1	7·5	9·5	12·1	14·9	16·3	16·4	14·8	11·9	8·6	6·7
Kiev	°C.	−6·1	−4·7	−0·2	7·2	15·2	18·8	20·2	19·1	14·7	7·7	1·3	−4·4

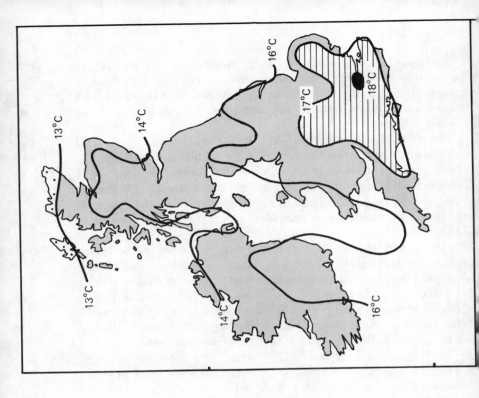
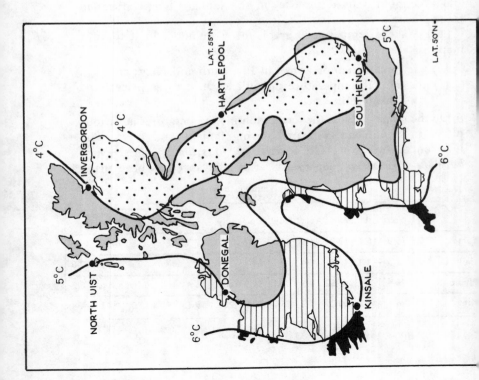

for altitude by 'reducing to sea-level'. This means adding, to the actual mean temperatures recorded, 1° C. for every 150 m a place lies above sea-level. For example, the *actual* mean January temperature at the summit of Ben Nevis is —3·9° C. As the mountain is 1343 m high $\frac{1343° \text{ C.}}{150}$, or 8·9° C. are added to give a converted 'sea-level' mean January temperature for Ben Nevis of 5° C.

(*20*) Work out the converted mean July 'sea-level' temperature at Buxton (335 m above sea-level in the Pennines), given that the actual July mean temperature there is 14·2° C.

(*21*) Now look carefully at maps A and B, and write out the following paragraphs correctly:—

IN JANUARY the isotherms tend to run $\frac{\text{NORTH-SOUTH}}{\text{EAST-WEST}}$. The $\frac{\text{SOUTH-EAST}}{\text{NORTH-WEST}}$ coast of Scotland is as mild as $\frac{\text{KENT}}{\text{CORNWALL}}$ at this time of the year. This is due to the influence of the $\frac{\text{COLD, DRY,}}{\text{WARM, MOIST,}}$ westerly airstream from over the $\frac{\text{WARM}}{\text{COLD}}$ North Atlantic Drift. Temperatures $\frac{\text{FALL}}{\text{RISE}}$ steadily towards the east coast, i.e. $\frac{\text{TOWARDS}}{\text{AWAY FROM}}$ the $\frac{\text{WARMING}}{\text{COOLING}}$ influence of the prevailing westerly winds. The isotherms bend $\frac{\text{SOUTH}}{\text{NORTH}}$ over the sea, showing the sea to be $\frac{\text{WARMER}}{\text{COOLER}}$ than the adjacent land in the winter.

IN JULY the isotherms tend to run $\frac{\text{NORTH-SOUTH}}{\text{EAST-WEST}}$. The altitude of the sun, and not the westerly airstream, is now the main factor affecting temperatures in Britain. Temperatures are high towards the $\frac{\text{SOUTH}}{\text{NORTH}}$ and $\frac{\text{SOUTH-EAST}}{\text{NORTH-WEST}}$; the London area is particularly $\frac{\text{WARM}}{\text{COOL}}$. The isotherms bend $\frac{\text{SOUTH}}{\text{NORTH}}$ over the sea: the land is now $\frac{\text{COOLER}}{\text{HOTTER}}$ than the adjacent seas.

The warmest and the coldest parts of the British Isles are differently shaded on each of the maps A and B. (*22*) Trace the maps, shade the warmest parts red and the coldest parts green, and add a labelled key to each map.

(*23*) Finally, use Map A to complete the notes below, which will then show clearly the remarkable warming influence of the westerly airstream on the west coast of the British Isles during the winter months:—

Latitude (Approx.)	West Coast Station	Mean Jan. Temp. ° C.	East Coast Station	Mean Jan. Temp. ° C.	West is warmer than East by
	Kinsale Donegal North Uist		Southend Hartlepool Invergordon		

1. Sun's heat causes water to evaporate in form of millions of _invisible_ water particles from:

OCEAN SURFACES

RIVERS   LAKES

VEGETATION

2. At a given temperature air continues to absorb water vapour until it becomes _saturated_.

3. When saturated air is cooled it can no longer contain so much water vapour: the surplus vapour _condenses_ to form masses of _visible_ water droplets, i.e. CLOUDS

4. The cloud droplets gradually link together to form larger water drops which in time fall as RAIN.

Rain falls so frequently in Britain that we rarely pause to consider how vitally we depend upon it. In fact water is in such great demand, for drinking, washing, sewage disposal, industrial uses and for growing crops that it is convenient to know exactly how much of it falls at any one place. Rainfall is measured by a **rain gauge.** This instrument consists of a funnel which guides into a can the rain falling on it. The rain gauge is placed in an open position, away from buildings and walls which would provide shelter from the rain. Every morning any water in the can is emptied into a cylinder which shows the amount of rain which has fallen in inches. The tapered end of the measuring cylinder enables the rainfall to be read correct to 0·05 mm. It shows the depth to which level ground would have been flooded if the rainwater had not drained away, evaporated or soaked into the ground.

The annual rainfall at any place is found by adding together all the daily amounts. As that year may have been unusually wet or dry, an average is taken of the totals for 30 or 40 years. The result is the **mean annual rainfall** for the place where the measurements have been made.

One inch of rainfall is a very large amount, and except during thunderstorms it is unusual in Britain to record so heavy a fall in one day. When such heavy amounts do fall, serious flooding is likely. The disastrous flooding at Lynmouth in August 1952 (see page 22) was the result of 225 mm of rain falling on Exmoor in one night.

The chilling of water-saturated air, which brings

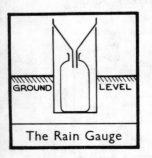

GROUND   LEVEL

The Rain Gauge

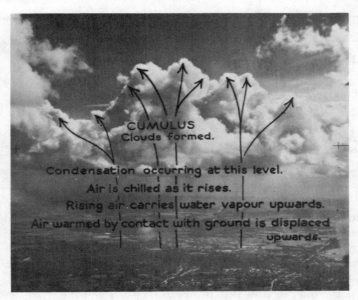

CUMULUS
Clouds formed.

Condensation occurring at this level.

Air is chilled as it rises.

Rising air carries water vapour upwards.

Air warmed by contact with ground is displaced upwards.

Clouds formed in this way have level bases (Why?), but billow upwards in great mushroom-shaped tufts. They are **cumulus** *clouds, from which heavy rain may fall, often accompanied by thunder. Man-made cumulus clouds are formed during nuclear explosions. Compare a photograph of such an explosion with the picture above. (24) Why do glider pilots steer their aircraft towards the base of cumulus clouds?*

about condensation, cloud formation and rainful, can take place in three main ways.

1. **Convection rainfall.** (*25*) Place a beaker of water on a tripod over a bunsen burner, and sprinkle a little sawdust into the water. Gently heat, and carefully watch the movements of the dust. You will see it circulate in the beaker, rising in the middle and falling near the edges due to the **convection** currents set up in the water by the heat at the bottom of the beaker. As the lower layer of water is warmed, it expands and becomes less dense. Cooler, denser water at the surface then sinks to the bottom, displacing the warmer water upwards. Exactly similar convection currents are set up in the air when the ground surface is strongly heated by the sun. The details are shown in the photograph.

Daily convection rain and thunderstorms are common in the tropics because of the intense heat. (*26*) Suggest reasons why such rain in Britain (*a*) falls mostly in summer; and (*b*) is most widespread in the East Midlands?

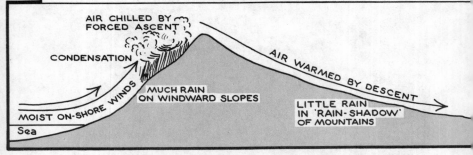

AIR CHILLED BY
FORCED ASCENT

CONDENSATION

MOIST ON-SHORE WINDS

MUCH RAIN
ON WINDWARD SLOPES

AIR WARMED BY DESCENT

LITTLE RAIN
IN 'RAIN-SHADOW'
OF MOUNTAINS

Sea

2. **Relief rainfall.** When moist air blows towards a hilly or mountainous coast it is forced to rise to cross over the high ground. The forced ascent chills the air, and the rainfall which results is called *relief rain*. Naturally, this type of rainfall is most common in those parts of the world where, as in west Britain, moist winds blow onshore throughout the year.

As the high land in Britain lies in the west and the north-west, these parts of the country receive the heaviest falls of relief rain. Very many tourists have returned home disappointed from the Lake District, North Wales or the Scottish Highlands, where rain and mist obscure the scenery for so many days of the year. Farther east in the lee of the mountains is a **rain shadow** zone where little relief rain falls. (See diagram.)

3. **Frontal rainfall.** Overleaf (page 59) is an isobar pattern often seen on air pressure maps of North-West Europe—the ' **depression** ' or '**low**'. Listeners to the B.B.C. ' weather forecasts ' know how frequently the announcer refers to a depression over Iceland, or off North-West Scotland, or perhaps off South-West Ireland. *A depression is an area of low atmospheric pressure*, in which air swirls around and towards the centre in an anticlockwise fashion. Cold, heavy air forces its way from the outer fringe towards the low-pressure centre.

The diagram overleaf shows that on its way it drives beneath warm, moist, lighter air, lifting the latter bodily upwards and lowering its temperature. The chilling causes water-vapour to condense, so that clouds form, and rain falls. Within most depressions there is a **warm sector,** where warm, moist air blows, trapped between cold air on each side. Such is the case between *O* and *P* on the map overleaf. (27) See if you can write the following account correctly:—

As the warm air moves forward it gradually slides $\frac{\text{UP OVER}}{\text{BELOW}}$ the $\frac{\text{LIGHTER}}{\text{DENSER}}$,

CIRRUS
(A high-altitude cloud)

CUMULUS

colder air, the boundary between the warm and cold air making a $\frac{\text{GENTLE}}{\text{STEEP}}$ angle with the ground surface. All along this invisible boundary, called the $\frac{\text{COLD FRONT}}{\text{WARM FRONT}}$, the warm moist air is $\frac{\text{CHILLED}}{\text{HEATED}}$, and clouds form.

At $\frac{\text{HIGH}}{\text{LOW}}$ altitudes the clouds are thin wispy threads called $\frac{\text{CUMULUS}}{\text{CIRRUS}}$: lower down come $\frac{\text{CIRRUS}}{\text{CUMULUS}}$ clouds, followed by $\frac{\text{CUMULO-NIMBUS}}{\text{CIRRUS}}$ from which steady rain falls. Where the warm air is still in contact with the ground, the sky is $\frac{\text{OVERCAST}}{\text{CLEAR}}$.

Towards the rear of the warm sector, $\frac{\text{COLD}}{\text{WARM}}$ air thrusts forward and lifts the $\frac{\text{COLD}}{\text{WARM}}$ air off the ground. Here, at the $\frac{\text{COLD FRONT}}{\text{WARM FRONT}}$, $\frac{\text{CUMULUS}}{\text{CIRRUS}}$ clouds develop and clearing showers fall. In time the $\frac{\text{WARM}}{\text{COLD}}$ air is lifted completely off the ground to form an **occlusion** (see Diagram B). Thick cloud cover usually marks an occlusion, and considerable rain is likely to fall from it.

A succession of depressions drift across the British Isles every year, reaching us from the North Atlantic and travelling in an easterly direction. They are largely responsible for the characteristic changeability of British weather (rain one day—fine the next). The pictures overleaf show how, during the passage of a depression, the weather may be very different in various parts of the country on the same day.

57

Weather at Falmouth. *Weather at Manchester.*

(28) Imagine you were a weather observer in Dover during the passage of the depression shown on this map. Using the information in the map, diagrams and pictures, write down the correct version of the following report:—

" The approach of the depression was heralded by a $\frac{RISING}{FALLING}$ barometer, $\frac{SOUTH\ TO\ SOUTH\text{-}WEST}{NORTH\ TO\ NORTH\text{-}WEST}$ winds, and a $\frac{DENSE}{FILMY}$ covering of $\frac{HIGH}{LOW}$ cloud. As the centre of the depression reached us the temperatures $\frac{FELL}{ROSE}$, and the barometer $\frac{ROSE}{FELL}$ still further, and the cloud cover became $\frac{LOWER}{HIGHER}$ and $\frac{THINNER}{THICKER}$: first $\frac{PROLONGED\ RAIN}{DRIZZLE}$ and then $\frac{DRIZZLE}{PROLONGED\ RAIN}$ fell. After the centre had passed the barometer $\frac{FELL}{ROSE}$ again, temperatures $\frac{DECREASED}{INCREASED}$ and the wind veered around to the $\frac{SOUTH\text{-}WEST}{NORTH\text{-}WEST}$. At the rear of the depression the wind was $\frac{FRESH}{GENTLE}$, visibility improved and there was occasional thunder."

(29) If you had been reporting the passage of the same depression from Aberdeen, what difference would you have noticed in the gradual change in direction of the wind?

Weather at Dover.

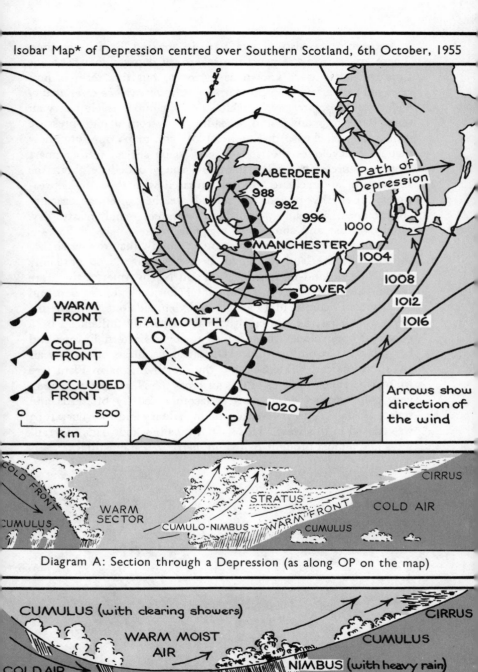

Isobar Map* of Depression centred over Southern Scotland, 6th October, 1955

ABERDEEN
988
992
996
1000
MANCHESTER
1004
DOVER
1008
1012
1016
FALMOUTH
O
1020
P
Path of Depression

WARM FRONT

COLD FRONT

OCCLUDED FRONT

0   500
km

Arrows show direction of the wind

COLD FRONT
CUMULUS
WARM SECTOR
CUMULO-NIMBUS
STRATUS
WARM FRONT
CUMULUS
CIRRUS
COLD AIR

Diagram A: Section through a Depression (as along OP on the map)

CUMULUS (with clearing showers)
WARM MOIST AIR
CIRRUS
CUMULUS
COLD AIR
NIMBUS (with heavy rain)

Diagram B: Section through an Occluded Front

*Based on weather map issued by the Director General, Meteorological Office

Anticyclones also play a part in making British weather. These are in many respects the reverse of depressions. (At one time the latter were known as cyclones, but this term is now reserved for certain tropical storms.) *An anticyclone is an area of high atmospheric pressure*, in which a cool, comparatively heavy air mass is settling down and spreading out from a high-pressure centre in a clockwise direction. *(30)* The maps opposite show that in anticyclones the isobars tend to be spaced farther apart than in depressions: what, therefore, can you deduce about the strength of anticyclonic winds, compared with those in depressions? Lengthy periods of settled weather often accompany anticyclones: hot, sunny and dry in summer—cold, frosty, dry, often overcast and sometimes foggy in winter.

In winter the land loses its heat more rapidly than the sea so that the air over the land is chilled, becomes denser, and gradually sinks. Great land-masses like Asia and North America therefore become high-pressure centres in winter; anticyclones develop, and bitterly cold 'continental' air gradually spreads out from the interior. During the winter Britain is frequently influenced by a powerful anticyclone which builds up over Eastern Europe and Scandinavia. Coming from Arctic latitudes, blasts of icy cold air sweep across the Baltic Sea and the North-European Plain (see Map *A*). When tongues of this air cross the North Sea to Britain very cold weather ensues, often accompanied by blizzards and severe frosts. A great 'freeze-up' in January and February 1963 was caused in this way. Heavy snowfalls blocked roads and railways and almost brought British industry to a standstill.

Traffic congestion due to snowy conditions on the trans-Pennine motorway (M62) near Leeds.

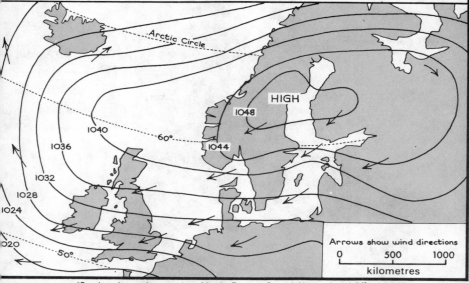

Map A: Isobar Map* of Typical Winter Anticyclone centred over Scandinavia.

HIGH
1048
1044
1040
1036
1032
1028
1024
1020

Arctic Circle
60°
50°

Arrows show wind directions

| 0 | 500 | 1000 |

kilometres

*Based on the weather maps issued by the Director General, Meteorological Office

In summer the sea takes longer to heat up than the land. Consequently the air over the sea tends to be comparatively cool. In the North Atlantic such air settles down over the Azores, and a marked anticyclone develops. Air spreading outwards from the 'Azores High' sometimes reaches Britain, bringing a gentle stream of sub-tropical air and a 'heat wave' to our shores from the south-west. (See Map *B*.)

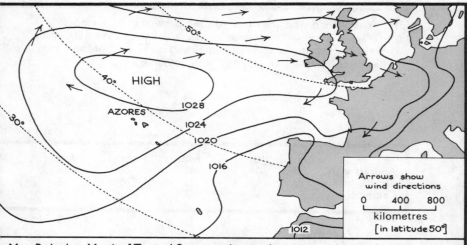

HIGH
AZORES
1028
1024
1020
1016
1012

50°
40°
30°

Arrows show wind directions

| 0 | 400 | 800 |

kilometres
[in latitude 50°]

Map B: Isobar Map* of Typical Summer Anticyclone centred over the Azores.

*Based on the weather maps issued by the Director General, Meteorological Office

British Isles : Rainfall

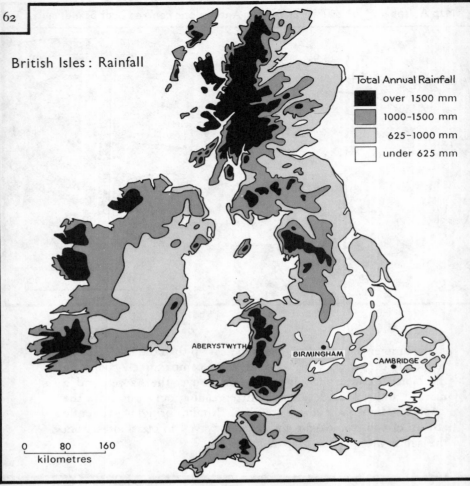

Total Annual Rainfall

- over 1500 mm
- 1000-1500 mm
- 625-1000 mm
- under 625 mm

ABERYSTWYTH

BIRMINGHAM

CAMBRIDGE

0   80   160
kilometres

(31) What broad similarity can you see between these maps?
(32) Trace the rainfall map, shading the areas with over 1500 mm.
Lay your tracing on the relief map, and make a list of the districts
where high relief and high rainfall coincide.

Mean Month

		Jan.	Feb.	Mar.	Apr.	M
Aberystwyth	. . .	104	80	88	66	6
Birmingham	. . .	52	43	49	44	5
Cambridge	. . .	38	32	37	34	4

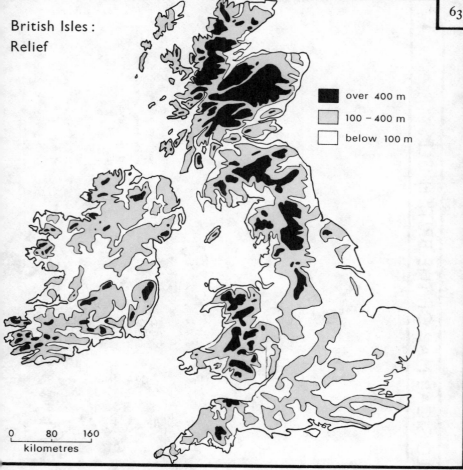

British Isles:
Relief

over 400 m

100 – 400 m

below 100 m

0   80   160
kilometres

(33) Make rainfall bar diagrams for Aberystwyth, Birmingham and Cambridge. Work out the total annual rainfall in each of these places and account for the differences. Can you explain the comparatively high amount of rain which falls at Cambridge in June, July and August? (Hint—re-read page 55.)

(34) Name the rainiest and the driest regions of (a) Ireland, (b) Scotland, (c) England and (d) Wales.

nfall (mm)

e	July	Aug.	Sept.	Oct.	Nov.	Dec.
	98	124	93	135	120	129
	59	69	46	71	61	69
	55	60	41	60	49	49

A SIMPLIFIED GEOLOGICAL MAP OF THE BRITISH ISLES

N.B. Many lowland parts of the British Isles are covered by Glacial Deposits.

	RECENT { RIVER ALLUVIUM, MARSH, ETC.
LC	LONDON CLAY AND SANDS
▦	CHALK
G	GAULT CLAY AND GREENSAND
W	WEALDEN CLAY AND SAND
J	JURASSIC
NRS	NEW RED SANDSTONE
	COAL MEASURES
C ⋰	CARBONIFEROUS LIMESTONE AND MILLSTONE GRIT
	ANCIENT SEDIMENTARY AND METAMORPHIC ROCKS
■	VOLCANIC ROCKS
+++	GRANITE ETC.

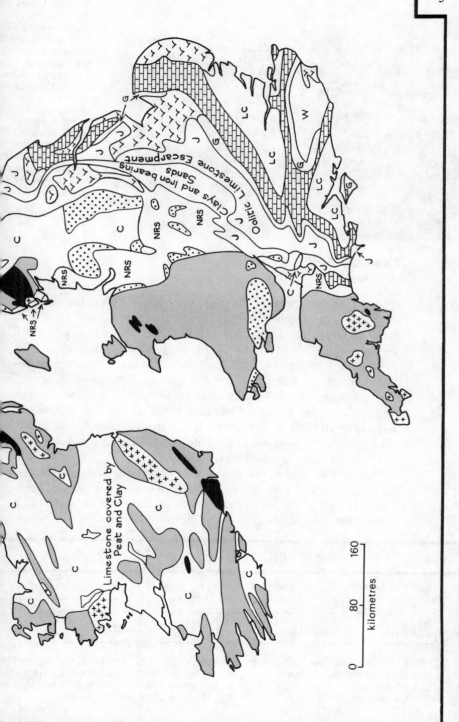

Oolitic Limestone Escarpment

Clays and Iron bearing Sands

Limestone covered by
Peat and Clay

NRS

0   80   160
kilometres

Desolate scenes li[ke]
this view of an
abandoned, roofles[s]
farmstead in the
Orkneys (map p. [ ]
have become
increasingly comm[on]
during the past tw[o]
hundred years,
throughout large [areas]
of the Scottish
Highlands and Isl[ands]
By contrast, the o[il]
storage tanks und[er]
construction in th[e]
background are p[art]
of a new and very
sudden change
(text p. 72).

CHAPTER 4
Scotland

How do we account for the remarkable difference between
these two scenes? What has discouraged men and their
families from remaining in some parts of Scotland, and drawn
them in hundreds of thousands to other localities near by? The
contrast is explained by the contrast in **natural resources** between the three different regions of Scotland.

These regions are clearly defined by two long faults which
mark off (*a*) the Central Lowlands from (*b*) the Highlands and
Islands to the north and (*c*) the Southern Uplands to the south.
(*1*) Using the map on page 79, shade and label a simple sketch map
of Scotland to illustrate the previous sentence.

During the Ice Age the North-west Highlands and Islands, the
Grampians and the Southern Uplands were all heavily glaciated.
Ice-caps and glaciers removed much of the existing soil. The
mountains show all the features of highland glaciation described on

WHERE PEOPLE LIVE IN SCOTLAND

Central Lowlands and Coast Plains	Highlands, Islands and Southern Uplands	1 Million People =
𝑿𝑿𝑿𝑿	𝑿	

NOTE Since 1960 the population of
Scotland has remained almost unchanged
in total numbers; but for the first time
in over a century the population of the
Highlands has increased slightly instea[d]
of falling. As elsewhere in Scotland, th[e]
increase affected the towns rather tha[n]
the country districts.

A busy street in the great ship-building port of Glasgow, a city with almost a million inhabitants —one fifth of all the people in Scotland.

pages 32–35 (*photo p. 34*). Huge tracts of the Scottish Highlands are almost bare of soil and are of little agricultural value. Another disadvantage of the Highlands for farming is the heavy rainfall, which is especially marked towards the west. ((2) *Why?*) Even when it is not raining or drizzling, many days are overcast, so that crops get little chance to ripen.

In the remote mountains and islands, particularly towards the north-west . . . " the hills are large and gaunt and threatening, for the most part, the habitations few and small and white. Britain has no wilder stuff to show; the urban dweller is at once affrighted, fascinated and delighted by these brown wildernesses, where there may be the risk of death in any casually planned walk across the moors."[1] Whether by road, rail or sea, communications are generally scanty, difficult and slow. Such a countryside may attract holiday-makers, but for permanent residents life is hard, expensive and restricted. People have been leaving the High-lands steadily over the past two hundred years for less harsh surroundings elsewhere in Britain and overseas.

In contrast to these bleak conditions the Central Lowlands, and to a lesser extent the narrow East Coast plains, had more to offer man. Here there were fertile soils and rich boulder clay which, together with the lighter rainfall, sunnier days and higher temperatures, favoured both arable and pastoral farming. More important, by the early 19th century, was the mineral wealth of the Central Lowlands. The iron ore and coal found there helped the building of important industries—iron and steel, shipbuilding, textile and engineering. It was the industries of central Scotland which attracted so much of the population into so small an area.

[1] George Blake (introduced by J. B. Priestley), *The Beauty of Britain* (Batsford).

0   25
kilometres

LAND OVER
200 metres

— — MAIN
RAILWAYS

Ferry

THURSO
WICK
INVERNESS
KYLE OF
LOCHALSH
ABERD
BALLATER
MALLAIG
Spey
Dee
FORT
WILLIAM
MONTROSE
Lochaline
Tay
Craignure
DUNDEE
OBAN
PERTH
STIRLING
DUNFERMLINE
EDINBURGH
GLASGOW
BER
SYMINGTON
GALASHIELS
Clyde
Tweed
AYR
Teviot
HAWICK
Nith
Annan
DUMFRIES
CARLISLE
LARNE
STRANRAER

(3a) Study these two pages and look again at pages 62–5. Can you suggest any links between the density of population in Scotland and (i) the geology or (ii) the relief or (iii) the rainfall? Explain your answers.

(3b) Draw a sketch map similar to that on page 84 to show the position of Glasgow as Scotland's principal rail transport **node** (centre). Label each arrow to

show where it leads.   Only the main r routes are marked; but in fact very f minor railways are now left in Scotla

(3c) In planning a railway route, t engineers must to some extent ta account of the relief. Point out fo examples from this map.

(3d) Suggest another, quite differe reason why there are more railways the lowlands.

Large tracts of *THE HIGHLANDS* are occupied by deer 'forests' and hill-sheep farms. These 'forests' contain few trees; the wind-swept mountain slopes are mainly covered by bare rock and scree (*p. 19*) or by coarse grass and mosses suitable only for rough grazing. Deer-stalking attracts some wealthy tourists. On the lower slopes, where soils are better and grass more abundant, cattle rearing replaces sheep farming.

In the remoter Western Highlands and Islands (including the Orkneys and Shetlands) a simple type of farming called **crofting** survives. The traditional crofter's home was a humble single-storied dwelling of rough-hewn stone with two or three rooms and a cow-shed adjoining. The crofter usually kept sheep, which wandered during most of the year over the neighbouring hills in search of rough herbage. He also owned one or two cows, some poultry and perhaps a few pigs. He grew almost all his own food, the staple being oatmeal. On the coast crofters added to the meagre produce of the land by fishing.

Today many crofts are abandoned (*photo p. 66*), their owners having left such a harsh life for easier conditions in the cities of Central Scotland. Yet 15 000 or so crofters still live in the Islands and all over the West and North. Modern dwellings, built with the aid of government grants, have replaced the former cottages. Most crofters now have a paid job—e.g. roadman, postman, forestry worker—and run their crofts as a spare-time activity. This fills in the hours after work, and counteracts the boredom which, in such quiet, remote places, may lead to a hankering for the excitements of town life. To encourage them to persist, despite the high cost of rearing livestock in this bleak countryside, the government pays crofters an annual **subsidy** for each breeding cow or ewe.

The croft provides at least potatoes and milk, and probably eggs and vegetables, together with a small profit each time a fat bullock or a few bales of wool are sold. By letting sites for caravans, by catering for 'bed and breakfast' and in various other ways many crofters and their wives benefit also from the rapidly-growing tourist trade.

Some crofters still make and sell hand-woven cloth. The famous 'Harris tweed' is a high-quality cloth made on hand-looms in the Outer Hebrides. At one time it was hand-spun as well as hand-woven, but today the spinning is done at several mills, mostly at Stornoway on the Island of Lewis. Though mass-produced cloth from modern factories is cheaper, there is a steady demand for genuine Harris tweed because of its very fine quality. The same is true of the traditional garments and shawls hand-knitted in the Shetland Islands.

Attempts have been made for many years past to develop new industries in the remote Highlands and Islands, and thus to encourage people to go on living there. Many such attempts have failed. Small, widely scattered settlements, mainly of elderly people, cannot provide a reliable supply of labour for modern industry. Most new enterprises have been based on the towns and larger villages, and the drift from outlying districts has continued.

(4) What other factors make the region unsuitable for most factory industries? (*Hint: raw materials? transport? markets?*)

One requirement for modern life and work—cheap electricity—is available almost everywhere in Northern Scotland, as explained opposite. This has made possible the establishment of aluminium smelters, including the one illustrated opposite.

Another example of new, large-scale industry in the Highlands is the pulp and paper mill opened near Fort William in 1966 on the site marked X. In the Highlands, as in other barren regions of Britain, wide areas of land have been planted with coniferous trees in the past fifty years. This ' crop ', now reaching maturity in rapidly increasing quantities, provides the raw material—1400 tonnes of softwood every day. The long loch offers sheltered, deep-water access for shipping. Every day the mill uses over 100 million litres of fresh water from the turbines of the nearby smelter's generators. Including 1500 forestry workers, the mill provides work for 2000 to 3000 men. It also saves the country millions of pounds each year, for most of Britain's pulp and paper has to be imported. (5) State, from the details given, what factors make the Western Highlands a suitable location for both the industries mentioned.

Unfortunately tree growth in this northerly latitude is slow, and it will never be possible to keep many such mills supplied with home-grown timber. There are also problems of air and water pollution. Probably the most rewarding industry for the Highlands will always be tourism. First made fashionable by Queen Victoria, the Highlands now draw two million visitors every year despite the unreliable weather. Most tourists come between mid-July and early September, though a winter sports centre, opened at Aviemore in 1966, has become very popular. The money they spend sustains many small, scattered communities that would otherwise disappear. (6) Suggest examples of jobs

...wn these pipes to the power-house ...low rush the waters of Loch Treig, the Western Highlands. The ...chaber hydro-electric power station ...rms part of this aluminium re-...ery near Fort William.

ELECTRICITY IN THE HIGHLANDS AND ISLANDS

...he first Scottish hydro-electric station ...built as long ago as 1896 at Foyers, on ...h Ness. The current was used for re-...g aluminium, a process which requires ...e supplies of clean, cheap power. ...he success of this venture led to the ...ding of two more aluminium refineries, ...at Fort William (above) and one at ...ochleven, each with its own hydro-...tric power plant; but there was still ...public electricity supply throughout ...Highlands.

...he North of Scotland Hydro-elec-...ity Board was set up in 1943 with the ...berate purpose of making domestic ...working life easier throughout the ...on, and thus of encouraging people to ...there instead of joining the 'drift to ...South'.

...hy hydro-electricity?

...lthough the Highlands are largely ...en and unproductive, they are well-...ed to the production of water power:

... There are many waterfalls, lochs and ...st-flowing torrents.

...) The building of dams is made easier ...y the narrow valleys of the glaciated ...ountains, with solid rock usually found ...ose beneath the surface.

...ii) The rainfall is heavy and reliable.

...v) Loss by evaporation from the reservoirs is small, for skies are often overcast and temperatures are moderate.

(v) Except on high ground the risk of freezing is slight. ((7) *Why is this an advantage?*)

Most people in the Highlands, even in quite remote districts, now have mains electricity. The Board has over 60 h.e.p. stations; but only 9 of them have over 40 MW capacity, and even the largest, the **pumped storage** plant at Cruachan, has only 400 MW—small compared with the giant 2000 and 3000 MW **thermal** plants built in other parts of Britain in the past few years. Nevertheless the Board's work has been most useful. Its output is now equivalent to about 2m tonnes of coal a year, or about 1% of Britain's total electricity production (*see also pp. 268–9*).

The Board's water-power resources are reinforced by six diesel-powered stations in the Islands. Extra power for the more densely populated east coast comes from a 240 MW oil-fired thermal station at Dundee and from a 250 MW nuclear station at Dounreay. A new 1320 MW plant, to burn either gas or oil, is being built at Peterhead (*map p. 268*).

partly or wholly dependent on the tourist industry.

Better prospects for industrial development are found on the narrow coastal lowlands of the north-east, where a fairly dense population depended until the 1970s on primary industries— farming, fishing and quarrying—and on the service trades, including tourism. *Aberdeen* (181 000) is an ancient city with long-established commerce and industry. *Invergordon* had under 2000 inhabitants until a large aluminium refinery was built there in 1971 to take advantage of the cheap land, deep-water access and cheap nuclear power from Dounreay.

In both places—as almost everywhere throughout north-east Scotland—life has now dramatically altered. By June 1974 a *Times* correspondent could write: " Just three years ago Aberdeen was best known to the outside world for fish, beef, ... jokes about thrift, and as the royal gateway to Balmoral. " Now, he went on, Aberdeen was " the offshore capital of Europe ". (8) Explain each of the five points in this quotation.

North Sea oil and gas (*map pp. 268–9*) have brought fortunes to some people in this corner of Scotland, prosperity to many more, and problems to everyone. The problems are perhaps most marked in a very remote and tiny community like the Shetland Islands, where a huge and well-paid (but temporary) work force is building a terminal to deal with half Britain's oil supplies.

New, large-scale industry, even in an established centre like Aberdeen, demands an enlarged **infrastructure**—better road, rail and air links, more shops, houses, schools, banks, doctors and so on—which cannot be provided quickly. Oilfield jobs offer higher wages than shops and garages—which close because their staff leave. Oil company staff arrive with money to spend— house prices rise to three and four times their former level. These, and other such difficulties, offset some of the benefits that North Sea oil is bringing to north-east Scotland. The main question, though, is how much of the new-found prosperity will continue beyond the 1990s, when the oil may cease to flow. For instance, will Stornoway, now employing a thousand men on rig repair work, or Peterhead, now building a big petro-chemical factory, once again become no more than small fishing ports subject to heavy seasonal unemployment?

(9) Suppose a major oilfield were discovered in the Firth of Clyde. Discuss, after reading pages 79–83, whether similar problems would arise in quite so marked a form.

FARMING AND FISHING IN NORTH-EAST SCOTLAND

ok again at the map on page 255. Why do the
␣er lands along Scotland's north-east coast show
␣h a contrast with the desolate Highlands?
␣he plain of Caithness and the sheltered plain
␣und the Moray Firth are composed partly of Old
␣ Sandstone, which gives a deeper, more fertile
␣ than the old, hard rocks of the Highlands.
␣tile boulder clays, too, are found on the lower
␣und. Since the prevailing winds drop much of
␣ir moisture on the Highlands ((10) Why?) the east
␣st plains receive an adequate but not excessive
␣an annual rainfall—about 750 mm.

␣lthough the climate is too cloudy and cool for
␣eat, malting barley is the main crop around the
␣ray Firth and over half of Scotland's malt whisky
␣ade in Morayshire. For the most part farming is
␣ixed'; grass, oats, seed potatoes and turnips are
␣ other leading crops. The rearing of
␣f cattle—the Aberdeen Angus breed is
␣rld-famous—is the most important
␣ivity, especially on the Buchan low-
␣ds, but dairy cattle and sheep are also

kept in large numbers.

(11) Using the map on page 68, draw a
simple cross-section across northern Scotland from Mallaig to Peterhead. Add the
following labels in appropriate places:

RTILE OLD RED SANDSTONE PREVAILING WESTERLIES BOULDER CLAY
D, HARD, INFERTILE ROCKS (draw arrows) HIGHLANDS
RY LITTLE FARMING MANY BEEF CATTLE BARLEY
KED FARMING SOME FORESTRY HEAVY RAINFALL
DDERATE RAINFALL BUCHAN LOWLANDS

Except for Inverness, all the coastal
␣wns shown on the map are fishing
␣rts. They are linked by road or rail
␣he Central Lowlands, where there is

a big demand for fish in the industrial
cities. Except for Grimsby and Hull
(page 141) more fish are landed at
Aberdeen than at any other British port.

(12) Inverness and Aberdeen are both important route centres.
Draw sketch-maps like that on page 84 to show the location of each
of these towns, using the map above, the notes below, and your
atlas.

Routes converge on Aberdeen from:—	Routes converge on Inverness from:—
Peterhead; Inverness;	Wick; Kyle of Lochalsh;
The Highlands (via Ballater);	Aberdeen; Fort William (via Glenmore);
The Central Lowlands (via Montrose);	Perth (via Tay and Spey Valleys).
Scandinavia; Germany.	

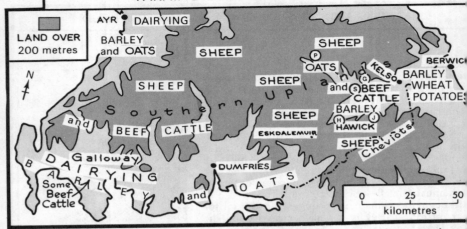

Notice on this map the wide coastal plain of Galloway and the smaller plain (known as the Merse) around Berwick. Both have very fertile soils and (for Scotland) a favourable climate.

From the information given on this page:

(13) Why does dairying represent 60% of all farming on the Plains of Galloway, whereas in the Tweed Basin cattle ar reared mainly for beef?

(14) Why are most sheep found in th eastern part of the Southern Uplands?

(15) Why are cereal crops more impor tant in the Tweed Basin than elsewhere?

(16) Why are barley and oats the mai crops of the Upper Tweed Basin, bu barley and wheat in the Merse?

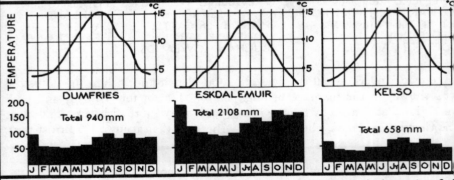

The main requirements of the crops and livestock mentioned on this map are as follows:

DAIRY CATTLE thrive best on moist clay plains and lower hill slopes, where grass is luxuriant. A mild climate is best, for the cows can then graze outdoors most of the year, thus lowering the cost of feeding.

BEEF CATTLE also need good pasture, but can be reared under drier and cooler conditions than dairy cattle.

SHEEP can be reared on moorland useless for other animals or crops. Upland breeds can withstand wet weather provided the pastures are well drained, but suffer various diseases on marshy ground.

WHEAT gives (in the right conditions) the highest yield and the highest price of al temperate cereals, and so farmers grow i wherever they can. In Britain, wheat grow best in a fertile and fairly heavy soil, wit plenty of sunshine and a mean annual rain fall of 550–750 mm.

Both barley and oats do well in simila conditions to wheat, but—

OATS thrives in moist, cool condition where wheat would not ripen. Oats make good livestock feed.

BARLEY can thrive in drier weather an on poorer soils than wheat. New type of barley also do well in moist, cool con ditions, and barley is now the leading cro in districts like Galloway, where oats wa once the only cereal.

This scene is a familiar one in the Southern Uplands, the most important sheep-rearing district in Britain.   Here the moorlands are lower and less rugged than in the Highlands, and pastures are better.   On the high moors the sheep (as in this photograph) are Black— face—a hardy breed able to withstand the frequent snow of the high plateaus and reared for their meat; Cheviot sheep, reared for their wool, are found on the better hill-lands and richer lower pastures.

SOUTHERN SCOTLAND is a plateau of ancient rocks, heavily eroded by ice and running water.   The rounded summits vary from 600 to 1000 metres above sea-level   They are separated by long, deep river valleys along which nestle the towns and villages and most of the scattered farms.   (*17*) Name five of the rivers from the map on page 68.

" The road rises out of flat, fat farming land, climbing among trees until, imperceptibly, the farms have become smaller, the rivers have dwindled to brown streams, growing crops given place to smooth bronze-green hillsides, marked here and there with a fan of scree thrown out by an occasional winter torrent. Then the deciduous trees disappear and there is only an occasional clump of fir or a plantation of spruce and pine.   Then the river is a little burn running in a deep ditch through the soft peat. Then it gets colder, and the gradient eases, and you may see mile upon mile of moorland, of which the feature is mainly its curious straw-brown colour, with a tarn here and there reflecting whatever colour may be in the sky and the cries of peesweeps and lambs emphasising the loneliness.

"These Uplands are notably given over to sheep farming.   It is the country of the gaunt, calm men who . . . know by headmark every sheep of a flock that may be 1000 strong." *

The largest number of sheep is found in the upper basin of the River Tweed, and it is not surprising that the Tweed Valley has long been famous for its woollen cloth industry.   At first the

* *The Beauty of Britain.*   (For a detailed study of a Southern Uplands sheep farm *see* Book 1 Chapter 2.)

Galashiels lies along a tributary of the Tweed called Gala Water. Factories both old and new line the river bank as shown here. Two main roads and a railway (now closed) follow the valley, but open country is near at hand on either side.

wool was spun and woven by hand in shepherds' cottages, or by monks in sheep-rearing monasteries. Later, woollen mills were built along the banks of the River Tweed and its tributaries, for these swift-flowing streams provided power to drive the machinery. But the Tweed Basin has no coal close at hand. When steam-driven machinery came into use the local woollen industry was unable to use it cheaply, and so the district was left behind by the enormous and rapid growth of woollen manufacturing in West Yorkshire (see pages 120–123).

Today the Tweed woollen towns concentrate on high-quality goods. Jedburgh has added rayon and clothing to woollen manu-facturing. Hawick specialises in knitwear, and Selkirk and Galashiels in woollens, including cloth for naval uniforms. Peebles is well known for its 'tweed' suitings. All these spec-ialised products are sold in Britain and are also exported to Europe and the U.S.A.

A major problem has arisen from the mechanisation of farming, which has left many men without alternative work. The woollen mills employ mainly women; and hundreds of men have moved away, taking with them their wives and daughters and so leaving

76

the mills short of labour.   The closure of most local railways, too, has not encouraged people to stay, and in recent years the loss of population has been faster than in the Highlands.

Electronics and other light industries have moved into the Tweed towns, offering jobs for both men and women; but most of the factories are branches of big firms, and are the first to suffer short-time working when trade is slack.   Many other workers in the Southern Uplands are employed in supplying various services (a) to tourists and (b) to farmers.

(*18*) Identify on the map (p. 74) the towns named above.

(*19*) Which of the following factors: *raw material, power, labour, markets*, do you think favoured the original growth of the Tweed Valley woollen industry?   (*20*) Which factors are still favourable? (*Remember that modern factories are not steam-powered.*)

In the fertile *CENTRAL LOWLANDS* both arable and pastoral farming are important.   (*21*) Make a *large* map of the region from page 79 and use the notes below to complete a labelled ' Farming Map of the Central Lowlands ' similar to that for Southern Scotland on page 74.

(*22*) Write down all the similarities you can find between the farming of the Central Lowlands and that of Southern Scotland.

(*23*) Using the notes on page 74, explain why the products of the west and east of the Lowlands are different.

FARMING IN THE CENTRAL LOWLANDS

In both west and east: SHEEP and BEEF CATTLE are reared on the isolated HILLS, often being brought down to the lowlands in winter.

In the west:

OATS, and more recently BARLEY, are of moderate importance, mainly for stockfeed.

GRASS flourishes, and DAIRYING is very important.   Well over a third of all Scotland's dairy cattle are in Lanark, Renfrew and Ayr. (24) *Why?* (*Hints: pp.* 52, 62, 69)

NEW POTATOES and MARKET GARDENING are important round Girvan, where the fertile but SANDY SOILS are easy to work and quick to warm up in springtime.

Half of Scotland's ORCHARDS and about a quarter of the MARKET GARDENS are in the Clyde Valley, partly because frost is so rare.

In the east:

There are specially FERTILE SOILS formed on Old Red Sandstone and with large patches of boulder clay.

The RICH ARABLE LAND on the lowlands of Fife and the Lothians, north and south of the Forth, is among the best in Scotland.   BARLEY, WHEAT, SUGAR BEET and POTATOES are leading crops. The potatoes are grown largely for sale at high prices to English farmers as ' seed '. This is because the cooler climate of Scotland reduces the risk of disease.

BEEF CATTLE are the main cash product, especially in the Vale of Strathmore; and DAIRYING is widespread, especially near towns.

Much SOFT FRUIT (berries, etc.) is grown in the sheltered Carse of Gowrie and Vale of Strathmore.

A SIX-YEAR ROTATION is typical: e.g. barley *or* oats / potatoes / wheat / sugar beet *or* turnips / barley / grass (for hay or grazing).

The district around Edinburgh is especially important for its MARKET GARDENS, DAIRY FARMS and PIGGERIES.

North–South Section through the Rift Valley of Central Scotland

Grampians   Southern Uplands

NORTHERN BOUNDARY FAULT   SOUTHERN BOUNDARY FAULT

Central Lowlands

CRYSTALLINE ROCKS

VOLCANIC ROCKS

COAL MEASURES

ANCIENT SEDIMENTS

OLD RED SANDSTONE

The Central Lowlands of Scotland form a rift valley, let down between the Grampians to the north and the Southern Uplands to the south. As a result the surface rocks are remnants of strata which once covered a much wider area, but which have been protected here from the severe erosion that stripped the highland regions.

Most of these rocks break down to form fertile soil, and they include several deposits of coal-bearing rocks.

(25) Re-read pages 14–16 and explain the origin of the volcanic hills shown on the map opposite. (26) What four towns lie approximately at the ends of the two fault scarp lines?

We have already learned something of the reasons why 80% of the Scottish population live in the Central Lowlands. **(27)** Copy the map opposite, and add the population shading from the map on page 255. Your finished map will show in greater detail where the four million people in the Central Lowlands have their homes. Notice on it the links between the distribution of population and the coalfields, estuaries and farming land.

Within the rift valley forming the Lowlands many districts are far from flat. A lowland 'gate' links the Firths of Forth and Clyde, but to the north and south are hills of volcanic rock. Few people live in these hills, for rough pasture covers them and only sheep farming is important.

Most people live in or near the coalfields and work in the manufacturing industries that have grown up there, or in the **service occupations** (transport, shop-keeping, teaching and so on) that employ so many people in any modern community. The biggest centre by far is Glasgow, bigger even than the ancient capital, Edinburgh (*p. 84*).

Glasgow grew up at the lowest bridging-point of the River Clyde as a market centre for the surrounding Lowlands. Its location was excellent for the purpose. Four important routeways converge on the city, all of them following lowland 'corridors'; from the Firth of Forth; from the upper Clyde valley; from Ayrshire; and from the lowlands around the mouth of the Clyde. **(28)** Using the map opposite, draw a sketch map like that on page 84, adding labelled arrows to illustrate the previous sentence.

In the early 19th century Glasgow grew very fast. The results,

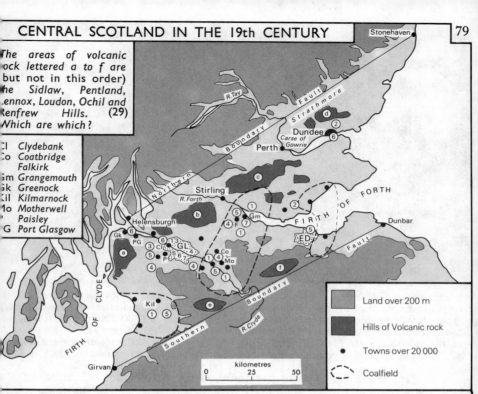

The areas of volcanic rock lettered a to f are (but not in this order) the Sidlaw, Pentland, Lennox, Loudon, Ochil and Renfrew Hills. (29) Which are which?

Cl Clydebank
Co Coatbridge
F Falkirk
Gm Grangemouth
Gk Greenock
Kil Kilmarnock
Mo Motherwell
P Paisley
PG Port Glasgow

Legend:
- Land over 200 m
- Hills of Volcanic rock
- • Towns over 20 000
- Coalfield

kilometres 0 25 50

In early times linen (from locally-grown flax) and woollens were widely made by hand. When steam power came into use the textile industry concentrated on the coalfields; WOOLLENS① in the Clyde Valley, in Ayrshire and near Alloa, and LINEN② in Fife and on Tayside. In the latter districts, both coastal, the manufacture of imported HEMP and JUTE (p. 85) was a natural development. Rather similarly, Glasgow's early trade with the American colonies led to the manufacture of imported tobacco, sugar and COTTON on the lower Clyde.③ (30) What factors favoured the location of each of these textile industries?

COAL MINING was unimportant until steam-engines were invented in the late 18th century. Coal then soon became the basis of Scotland's new prosperity, and also an important export cargo. (31) Four coalfields, the Ayrshire, Lanarkshire, Fife and Midlothian, are marked; which is which? (32) How did their location favour the export trade?

METALS had been worked by hand for centuries before 1801, when rich veins of 'blackband' iron ore were discovered in the Coal Measures. The occurrence of ore and coal together led to the growth of an IRON and STEEL industry④ in the early 19th century. By the 1850s Scotland produced one quarter of all British iron, but this proportion fell away when the 'blackband' ores gave out in the 1880s. Imported ores, however, allowed an enormous expansion of steel and ENGINEERING industries⑤ in Glasgow and in coalfield towns like Falkirk, Motherwell and Kilmarnock. (33) How did the shape of the Lowlands help to cheapen the import of bulk cargoes such as iron ore?

Of the engineering industries, SHIPBUILDING⑥ was outstanding. There were many centres, large and small, but the lower Clyde towns, Glasgow, Greenock and Clydebank, were world-famous. By the late 19th century a fifth of the whole world's ships were built on the Clyde. (34) Explain this, using the phrases: *early start / local coal and ore / skilled labour / deep, sheltered access.*

As usual in textile, coal and metal-working districts, CHEMICALS⑦ became important, especially at Glasgow and Grangemouth.

Warehouses and 'high rise' blocks of offices and flats have by now replaced most of the slums mentioned below. Factories and people alike have largely moved out to estates in the suburbs, or to 'New Towns' (map p. 83) specially planned and built to relieve overcrowding in Glasgow and other o[l] centres. The population of Glasgow itsel[f] which reached 1 100 000 in 1951, is expecte[d] to be about 750 000 in 1980.

Here are traffic-free 'walk-ways' and a pla[y] area among houses in Cumbernauld New Tow[n]

for those who had to live there, were not pleasant. Until only a few years ago "the oldest parts of the city ... had the highest density of population anywhere in Western Europe, containing nearly 100 000 slum homes and shops, 'dark-satanic-mill' factories, cobbled streets and alleyways, with all the filth, squalor and stench—not to mention crime—surviving from the first Industrial Revolution" (*Financial Times*).

Despite the bad living conditions, industry itself flourished. (*35*) Study list A and suggest why the adjective 'heavy' is used. (*36*) Which manufactures in list B are linked with Glasgow's function as a port?

Glasgow is still a great port and Britain's third largest city. Throughout the Lowlands, all the 19th-century occupations mentioned on page 79 remain important in varying degrees. Once industries become established they tend to continue in the same location even though the original conditions which favoured their growth no longer apply. This state of affairs is called **geographical inertia**, and it applies particularly to industries like steel and heavy engineering in which skilled labour and costly fixed equipment

A *Typical products of*		
HEAVY ENGINEERING IN THE LOWLANDS		
ships	boilers	earth-moving machinery
pumps	cranes	heavy electrical machinery
ships' engines		railway equipment
machine tools		mining equipment

B *Some other products of*	
GLASGOW and CLYDESID[E]	
flour	cigarettes
soap	cotton goods
paper	chemicals
petrol	rubber

This giant oil tanker is being escorted by tugs on its way up Loch Long to Clydeport's main oil terminal at Finnart (map p. 83).

The increasing size of tankers and other 'bulk carriers' has given a much greater value to deep, sheltered inlets like this.

play a large part. (37) Explain this sentence.

Nevertheless there have been big changes in the way many Scots earn their living, as shown by figures for the 1960s alone. Though farming, for example, is more productive than ever before, mechanisation has drastically cut the need for farm workers (1960—101 000; 1970—62 000). Coal-mining needs fewer men for the same reason, and also because many parts of the Scottish coalfields are now 'worked out' (1960—92 000; 1970—31 000). Both ship-building (1960—69 000; 1970—45 000) and textiles employ fewer workers, partly because of mechanisation and partly because foreign competition has taken much of their markets. This trend has continued throughout the 1970s.

One result has been more unemployment and a lower standard of living in Scotland than in Britain as a whole. Another result has been a serious attempt to get new industries into central Scotland, as outlined on page 83. Problems arise from the position of Scotland: isolated, like Northern Ireland, from Britain's 'centre of gravity' in the prosperous London/Midlands region, and isolated still more from the Continent.

THE GROWTH OF CLYDEPORT

Glasgow itself is a man-made port. Until the 19th century the Clyde at Glasgow was shallow enough to be forded at low tide. Ships had to unload at Port Glasgow, 29 km downstream (see map p. 79.) By dredging and blasting, a channel was made deep enough (8 m at low water) to allow most ships to reach Glasgow itself. Until recently the biggest of ships, including the 'Queen' liners, could be built and launched on the Clyde.

During the 1960s much bigger ships came into service. Fortunately the Firth of Clyde, formerly a glacial valley, very deep and steep-sided. All its

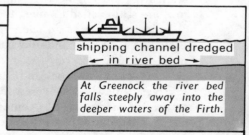

shipping channel dredged
← in river bed →

At Greenock the river bed falls steeply away into the deeper waters of the Firth.

ports are now controlled by one authority and called 'Clydeport'. The biggest of modern ships can unload somewhere in the port's area, and many people think Clydeport has a great future (text overleaf and map pp 282–3).

Some planners believe that Scotland's isolation may soon be ended by the growth in size of merchant ships. Some vessels already being built are too big for the shallow North Sea but not for the Firth of Clyde. An exciting concept called ' Oceanspan ' visualises Clydeport—claimed to be ' the best port in Western Europe '—as a sort of funnel into which the outside world would pour its bulk cargoes of oil, ore, coking coal, timber and so on. At Grangemouth and Leith smaller ships would reload for delivery to ports around the North Sea. On their way across the ' wasp waist ' of Central Scotland, some at least of these raw materials could be fully or partly manufactured, thus providing much-needed jobs for Scotsmen.

" *Oceanspan* is more than just a huge port complex . . . or even a ' land bridge '. It is an attempt to carve for Scotland a significant role in the pattern of industrial development within the Europe of the next couple of decades. Such a scheme . . . could trigger off activity comparable only to the first Industrial Revolution ". (*Financial Times*)

Other optimists claim that the newly-discovered offshore oilfields will make Scotland ' the Texas of Europe '. (*38*) What do they mean by this?

Current developments at Aberdeen or Peterhead, or in the Orkney and Shetland Islands, might seem to support this optimism. The most striking case, perhaps, is Cromarty Firth; on Nigg Bay, near the Invergordon aluminium refinery (*p. 72*), there is now a huge dry dock where oil production platforms are built; and a new oil refinery may soon be processing 10m tonnes of North Sea oil each year.

What of Central Scotland, though, where old industries have declined and unemployment and poverty are rife? Here, too, North Sea oil and gas have brought valuable work to many shipyards and engineering and chemical works; but many more jobs are still needed to offset those lost in the 1960s alone.

Industry depends on energy. In the 19th century this meant coal. Central Scotland now has big new power stations using not only coal but also oil, gas and nuclear fuel (*maps opposite and pp. 268-9*). The coal industry, modernised and with ample reserves of newly-discovered seams to mine, could have a settled future for many years to come; and much more offshore oil and gas may be found. Many people believe that Central Scotland has a real chance of regaining its former prosperity.

This map shows some new features of Central Lowlands industry in the late 20th century. It is not a complete map of the present distribution of industry, because the former pattern (map p. 79) largely survives in changed but still recognisable form.

D

North Sea oil and gas

Perth

Glenrothes

Finnart (oil terminal)

Stirling

Longannet

OIL PIPELINE

Leith

Gm

Cumbernauld

ED

Cockenzie

Gk

GL

Bathgate X

Livingston

Linwood X

East Kilbride

Hunterston
Ardrossan
Irvine

to England via M6

Bulk cargoes for Clydeport (p.82)

kilometres
0   25   50

▨	Land over 200 m
●	Towns over 20 000 (as p. 79)
o	"New Towns" (named on map)
▲	Thermal ⎫ power stations
△	Nuclear ⎭ over 1000 kw
✳	Very large coal mines
▢	"Container" ports and depots
▪	"Freightliner" rail terminal
▬	Motorways built or planned

(for main railways, see p. 68)

More than half Scotland's COAL now comes from only seven large mines (*see map*), and goes to power stations or to the STEELworks at Ravenscraig (near Motherwell). Steelmaking is likely to expand. A new unloading dock for giant iron-ore carriers opened in 1976 at Hunterston, where a big new steelworks is proposed for an adjoining site.

Industry is no longer tied to the coalfields, except by geographical inertia, for ELECTRICITY is available everywhere, generated mainly by very large new thermal and nuclear stations. (*39*) Name three of these. (*40*) Why were they located on the coast? (*Hint: NOT for access by sea.*)

Using the deep waters of the Firth of Clyde, supertankers deliver PETROLEUM in bulk to a terminal ((*41*) *Where?*) from which it is transported ((*42*) *how?*) to refineries at ((*43*) *what port?*) These refineries can treat 9 million tonnes of crude oil a year. They are being further enlarged to deal with oil and gas brought by tanker and pipeline from the offshore fields. Grangemouth is already an important centre of the PETROCHEMICAL industry, a natural development of oil and gas refining.

Except for MOTOR VEHICLE assembly plants at Linwood and Bathgate, most new factories are engaged in LIGHT INDUSTRY. They produce such things as computers, books, vacuum cleaners, neon lights, razor blades, underclothes and steel wool— a contrast with list A on page 80. Employment in ELECTRONICS alone rose from 7000 to 40 000 during the 1960s. Such factories are mostly on 'trading estates' or in NEW TOWNS (*caption p. 80*).

(*44*) What does the map tell us about attempts to meet the need for better LAND COMMUNICATIONS?

The Central Lowlands became a great industrial region because skilled labour, fuel and raw materials were all available within the region. (*45*) Which factors have changed? Illustrate your answer by referring in detail to two 'old' and two 'new' industries.

(*46*) Copy the map and add labels to summarise the information in these notes.

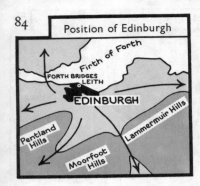

Position of Edinburgh

The dominating position of Edinburgh Castle —on the solid plug of an extinct volcano—is a constant reminder that the city once guarded the east-coast routeway into Scotland. Its position is still important, though now the routes towards the city are followed by road and rail traffic instead of by invading warriors.

Those shown on the map lead from Edinburgh to the following places: Glasgow, Perth, Ayrshire, Berwick on Tweed, Galashiels and the Tweed Valley. (47) Make a larger-scale copy of the map and, using an atlas, label each route correctly.

Edinburgh (472 000) is the capital of Scotland. Numerous government offices and the headquarters of many Scottish banks, insurance companies and other business firms are there, and it also has a famous university and several technical colleges and other educational centres, besides many schools. Law, catering and tourism are other ' industries ' which employ many people in Edinburgh because of the city's special position as capital.

To the north, Edinburgh has extended its boundaries to include the two seaports of Leith and Granton. Foodstuffs, chiefly grain and dairy produce, are imported. Oil, timber, scrap steel and coal are also handled, and fertilisers are manufactured.

MAIN INDUSTRIES IN EDINBURGH

Industry	Comments
Paper-making Printing Book-binding	These industries were stimulated by the demand for paper and books for the government offices and the university and law courts.
Engineering	Including scientific instrument making and electronics.
Brewing Distilling	(48) What crops used in these industries are grown nearby? Scotch whisky is one of Britain's chief ' dollar exports '. Edinburgh is a leading centre of the industry, which is Scotland's fifth largest employer.
Confectionery Biscuit making Food processing	These foodstuffs and those imported through Leith supply the needs of people in the industrial towns farther west, as well as in Edinburgh.

Dundee (185 000) is Scotland's third largest city and the focal point of Tayside, a small but distinctive sub-region stretching from Perth to the sea.

Its main industries used to be summed up as 'jam, jute and journalism', but this description is now incomplete and outdated, as shown by the notes. Journalism is still important and is typical of the service occupations mentioned on page 78.
(*49*) Which of Dundee's manufacturing industries do you think

---

SOME IMPORTANT INDUSTRIES IN DUNDEE

Manufacture of Linen, Hemp and Jute. Linen weaving originally grew up using local supplies of flax. Imports of hemp (a coarser fibre) were added as the North Sea fishing fleet increased its demand for sailcloth, ropes and nets. Later still (about 1830) jute was imported from India and manufactured into sacking, and this industry soon completely overshadowed flax and hemp manufacture. In recent years the jute industry in Dundee has suffered as new factories have opened in India and Bangladesh, where most of the world's jute is grown. Carpets and the backing for linoleum are now the mainstay of the Dundee mills, and man-made fibres are increasingly used in place of the (now more expensive) jute. New machinery has made the mills more efficient, but fewer workers are needed to operate it.

Making of Jam and Marmalade. Much soft fruit used in the jam factories comes from the nearby Carse of Gowrie (p. 77). Oranges for marmalade are imported from Spain.

Engineering includes cash registers, clocks and watches and cameras as well as machinery for the local jute and jam factories. This industry is now the main employer. One single branch of modern engineering—electronics—has more than balanced the loss of jobs in the jute mills.

Shipbuilding. Dundee ranks eighth amongst British shipbuilding ports.

Brewing. Barley of malting quality is grown in the locality.

Jute being processed in a Dundee mill.

grew up because of Dundee's position (a) as a port on a sheltered, deep-water estuary? (b) as the main market town of a densely-peopled, sheltered, south-facing belt of fertile farming land?

Dundee trebled its population in the late 19th century. In the 20th century it has been less prosperous. Its isolated location and the decline of the jute industry have held it back. In recent years new types of industry and a new road bridge southwards across the Tay have brought brighter prospects. The supplying and servicing of North Sea oil rigs has brought valuable trade to the port and repair work to the engineering firms.

The Highland Regiments of the British Army have a splendid fighting reputation. Their warlike traditions date back to medieval times when roving bands of highlanders made frequent raids to plunder the richer lowlands of corn and cattle. The easiest routes of attack lay through breaks in the line of volcanic hills to the north of the Clyde-Forth plain. In these gaps the lowlanders built defensive fortresses around which people settled for protection. Thus grew up the **gap towns** of Stirling and Perth. (50) From the map on page 79 write down which of the following gaps these fortresses defended:—

Gap A between the Sidlaw Hills and the Ochil Hills.
Gap B between the Lennox Hills and the Ochil Hills.

Stirling and Perth were also **route centres** and markets for the surrounding countryside. Later, railways and motor roads converged on them. In this respect both towns had additional importance, for they were at the **lowest bridging point** of wide estuaries. Today, however, the main road and railway both run due north from Edinburgh, crossing the Firths of Forth and Tay by bridges far to the seaward of Stirling and Perth.

" The three main railway routes into Scotland . . . traverse most . . . of the characteristic types of country to be found in the region. There is certainly a mighty difference between the process of getting to Edinburgh by the (old) London and North Eastern along the cliffy selvedge of the North Sea and that of reaching the capital with the aid of the (former) London, Midland and Scottish, cutting by Beattock and Symington through a tangle of bronze hills which, but for the seagulls on an occasional patch of ploughed land, might be a thousand miles from the coast. Then the western branch of the . . . old Glasgow and South

This view shows the Forth road bridge and (foreground) the approach to the famous railway bridge, both near Edinburgh. To reach the Fifeshire coast across the river from the capital it was previously necessary to go by ferry or to make a detour of 80 kilometres. Such bridges remind us how man by his ingenuity is able to surmount nature's obstacles.

Western line takes you up the valley of the Nith into Ayrshire through gentle and wooded country for the most part." (*The Beauty of Britain*, op. cit.)

(51) With the help of your atlas and the map on page 68, draw a large map to show the three railway routes into Scotland mentioned in the above account. Shade in the higher land, name the river valleys, and add the important towns on each route. (52) Then describe, and explain *fully*, what contrasts in scenery and general human activities you would expect to notice on journeys to Edinburgh (*a*) via Newcastle-on-Tyne and the East Coast Plain, and (*b*) via Carlisle and Symington. (53) Write a similar account contrasting road journeys from Carlisle to Glasgow via (*c*) the Nith Valley, and (*d*) the West Coast Plains.

(54) Suggest, with reasons, parts of Northern Scotland to which these descriptions might well apply:—

Place A " . . . these harsh coasts are backed by farming lands of a peculiarly rich quality. The fields run back from the edge of the cliffs to the hills, green and fruitful, the characteristic white walls of the Scots farmsteadings making delicious splashes against the monotone of pasture or ploughed land." (*The Beauty of Britain*, op. cit.)	*Place B* " . . . the maps show how fantastically it is indented by the sea, and the imagination should be able to conceive how these long and tortuous fiords add magnificently to the scenic effects. . . . Blue water, golden weed, green bracken on the lower slopes, and the brown moorland or the purple of heather above. . . ." (*The Beauty of Britain*, op. cit.)

CHAPTER 5

Wales

Gwlad, gwlad, pleidiol wyf i'm gwlad,
Tra môr yn fur
I'r bur hoff bau
O bydded i'r hen iaith barhau.

SOME OF you will probably recognise this as Welsh, but to most it will be completely meaningless. Does it surprise you to know that over 540 000 people in Wales can speak this language, and that of these over 10 000 cannot speak English? Did you know that many of the B.B.C. Welsh radio and television programmes are entirely in Welsh? How do we explain the survival of this ancient Celtic tongue? The answer lies largely in the geography of Wales.

Like much of west and north Britain, Wales is for the most part a mountainous country. Celtic-speaking tribes, who were once settled over much of Britain, took refuge in these mountains when attacked by Roman, Saxon, and finally Norman invaders. Secure in their mountain fortress home the Welsh retained their language, customs and nationality. Today the Welsh-speaking districts are mainly the remote villages and farms in the centre and north of the country. Attackers from England made little headway beyond the foothills. Notice how closely the present boundary follows the edge of the highland.

Wales: County Boundaries and Distribution of Population

Railway from	Notes
Chester to Holyhead Shrewsbury to Pwllheli Cardiff to Fishguard Shrewsbury to Swansea	Crosses the Plains of Pembroke; Passes nearest to a 1085-metre-high mountain; Crosses an area of Carboniferous rocks; Crosses a narrow sea-strait; Runs along the valleys of the Rivers Severn and Dovey; Passes nearest to the Brecon Beacons; Follows a narrow coast plain on Cardigan Bay.

Travel is still difficult in Wales. The mountains make an effective barrier to easy movement from north to south. Notice how the railways cling to the valleys and the narrow patches of lowland. (*1*) From the map below and your atlas sort out which of the notes in the above table apply to the four railway routes given.

Although the mountains give protection, life there has always been a struggle against nature. Many of the difficulties facing man in his attempt to wrest a living from the Scottish Highlands (see page 70) apply equally well here. Compare the population map (left) with that for Scotland on page 69. These similarities are important :—

1. In Wales, as in Scotland, most people live in the valleys or on the few plains.

2. The coalfield districts are densely populated.

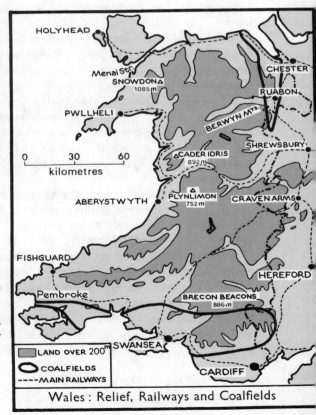

Wales : Relief, Railways and Coalfields

Here we see some of the 4 million sheep which form the chief wealth of Welsh farmers. Soils are thin on the windswept, glaciated uplands, temperatures are low (*Why?*), and it is often damp and misty. (Look at the climate chart for Rhayader.) Arable farming is out of the question, but sheep feed on the covering of heather and coarse grasses. On high ground this is possible only in summer, when the snows have melted. The sheep are driven up to the mountain pastures in the spring, and brought down to the warmth and shelter of the valleys for the winter.

Although soils are deeper and temperatures higher on the lower ground, drainage is often poor. Here sheep give way to cattle, for sheep suffer from foot-rot and other diseases on boggy pastures. The coastal lowlands of Anglesey, Pembroke and Glamorgan have an equable climate, due to the prevailing westerly winds. Summers are cool, winters mild, and the rainfall substantial. (2) (From the climate charts work out the annual temperature range for Haverfordwest and Rhayader.) Although barley and root crops are grown here, grass is lush and cattle are of greater importance.

Both beef and dairy cattle are reared. Dairy cattle are favoured on farms where (*i*) cows can be easily rounded up for milking, and (*ii*) the milk can be easily marketed. Beef cattle are preferred on remote, hilly farms. Intensive dairy farming in the Vale of Glamorgan supplies the needs of the industrial population of the South Wales coalfield.

For many years younger people have left the uplands of Wales to seek better paid jobs elsewhere. To counter this depopulation the Corporations for Mid-Wales Development and for Rural Wales aim to improve agriculture and forestry and to promote tourism and other industries. So far nearly £10 million has been

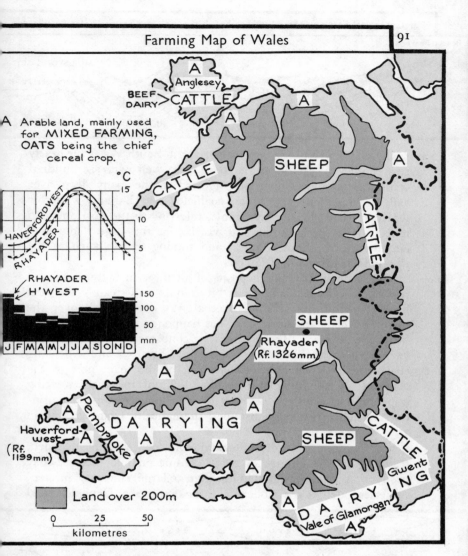

A Arable land, mainly used for MIXED FARMING, OATS being the chief cereal crop.

BEEF DAIRY > CATTLE
Anglesey A
CATTLE
CATTLE
SHEEP
A
A
CATTLE
A
SHEEP
Rhayader (Rf. 1326mm)
A
A
DAIRYING
A
A
A
SHEEP
CATTLE
Pembroke
Haverford-west (Rf. 1199mm)
A
A
A
A
A
Gwent
DAIRYING
Vale of Glamorgan
A

°C
15
10
5
HAVERFORDWEST
RHAYADER
RHAYADER
H'WEST
150
100
50
mm
J F M A M J J A S O N D

Land over 200m

0   25   50
kilometres

spent, mainly on laying out factory estates and building houses for employees at such 'growth points' as Newtown and Brecon. Cheap rents and other financial inducements are offered to firms willing to move in. Although some 50 factories have been established, providing a wide range of industrial jobs, there is still a steady drain of young people moving out of country districts. This exit is one of the main reasons for the decline of the Welsh language.

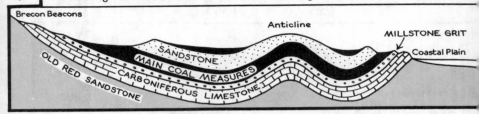

COALMINING has been important in South Wales for nearly two centuries. At one time there were well over six hundred collieries, most of which were located in the many deep river valleys which cut across the coalfield. (3) Explain why this location was especially favourable, referring to the diagrams and map. At its peak coalmining was by far the largest source of employment in South Wales, and mining valleys such as the Rhondda became world-famous.

The coalfield contains deposits of all three main types of coal (*see map*). Great quantities of bituminous and steam coals were formerly produced and Welsh coal was exported all over the world. The chart shows what has happened to this export trade in recent decades. Today mining is concentrated on (i) the very valuable anthracite seams to the north of Swansea and (ii) coking coal deposits along the southern rim of the main part of the coalfield. Coalmining still employs 4% of the male working population of South Wales, but it is no longer the principal industry, numbers having dropped dramatically from 101 000 in 1957 to 30 000 in 1978.

For decades the total demand for coal steadily declined, resulting in the closure of some 90 uneconomic pits. Today production is concentrated in less than 50 large, efficient, 'master' collieries, such as the modern anthracite mines at Cynheidre and

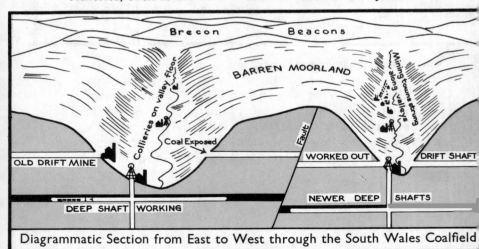

Diagrammatic Section from East to West through the South Wales Coalfield

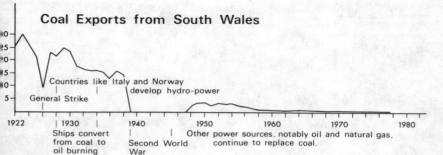

Countries like Italy and Norway
develop hydro-power

General Strike

1922 | 1930 | 1940 | 1950 | 1960 | 1970 | 1980

Ships convert
from coal to
oil burning

Second World
War

Other power sources, notably oil and natural gas,
continue to replace coal.

...uth Wales has always provided 30%–40% ...Britain's total coal exports. Thus the trade ...cline shown in this chart had a tragic ...pact on Welsh valley mining communities. ...e 1930s, in particular, were years of ...smal depression, unemployment and migra-...n. In recent years efforts have been made ...attract new light industries into the valleys. At Abercarn and Abertillery, for example, huge spoil heaps have been flattened into terraces on which new schools, houses, playing fields and some factories have been built. Most new jobs, however, are in industrial estates located at the valley mouths: see Q. 5, p. 96.

Abernant. Since the oil crisis of 1973–4 the decline in coal-mining has been halted: development is now the aim, with new investment and a recruiting drive to attract young men into the pits. Both the anthracite and the coking coal of South Wales are now assured of a long-term demand. The anthracite—Britain's only deposit—is the best in the world, and the coking coal forms 70% of U.K. reserves. Emphasis is being placed on opening new drift shafts and opencast workings, and in 1978 a large new anthracite mine was opened at Betws. 90% of coal in South Wales is now cut, loaded and conveyed by machinery, but costs are high because of the many faults and folds in the coal seams.

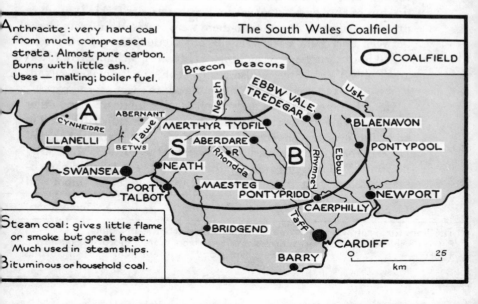

Anthracite: very hard coal from much compressed strata. Almost pure carbon. Burns with little ash. Uses — malting; boiler fuel.

The South Wales Coalfield

COALFIELD

Steam coal: gives little flame or smoke but great heat. Much used in steamships.

Bituminous or household coal.

Cooling molten strip steel at Margam Steelworks near Port Talbot. This is the largest steelworks in Europe, producing $\frac{1}{12}$ of all Britain's steel.

STEELMAKING. As early as the 14th century 'black-band' iron ore was found in limestone rocks in the northern part of the coalfield. Ore, limestone and fuel—first wood, later coal—were thus all readily available, and a string of iron-working towns grew up, such as Blaenavon, Ebbw Vale, Tredegar, Dowlais, Merthyr Tydfil and Aberdare.

In time, the best iron deposits were exhausted, but since then a vast industry has been built up using imported ore. Today all but one of South Wales' largest steel mills are on the coast, in or near Llanelli, Port Talbot and Newport (*map overleaf*). These works are well placed to receive iron ore from overseas, and trains bringing coking coal downhill from the nearby coalfield.

Much strip-steel (*see photo*) is produced in South Wales for the manufacturers of *tin-plate*. A 'tin' can is not made of pure tin, but of thin sheet-steel covered with a veneer ('skin') of tin to prevent rusting. At one time there were scores of tin-plate works between Port Talbot and Llanelli, but production is now concentrated in three large mills at Trostre (Llanelli), Velindre (Swansea) and Ebbw Vale. Steel is also coated with zinc, to make 'galvanised iron'—and a mixture of tin and lead, to make 'terne plates'.

In addition to the steel and tin-plate mills there are many other metal-working industries in the Swansea-Llanelli district. Tin and zinc ores are smelted, copper goods made, and the Mond Nickel Works near Swansea is the largest in the world. Metallurgy in South Wales is growing rapidly. (*4*) Draw a labelled sketch-map to show that (i) steel-making capacity at Port Talbot and

Tin-plate	'Galvanised iron'	'Terne plates'
'Tin' cans for: Fruit, Meat, Petrol, Fish, Vegetables, Tobacco, etc.	Roofing and fencing in tropical countries. (Why is iron more suitable than wood in these countries?)	Roofing, Motor-car bodies, Oil and paint drums

The new dock at Port Talbot, where a *ecial deep-water jetty has been dredged to *ke the world's biggest iron-ore carriers. *uch of the imported ore is consumed in the adjacent basic oxygen steel-making plant, newly built to replace old fashioned open-hearth furnaces.

Llanwern (Newport) is being expanded to 6m tonnes and 3.5m tonnes, respectively; (ii) Ebbw Vale has new steel-rolling, tin-plate and galvanised sheet works; (iii) a tube-mill is being built near Tredegar and (iv) an electric arc 'mini-mill' at Cardiff will produce steel to make rods and bars.

Steel-making is now the 'key' industry in South Wales, with one-sixth of the male workforce in steel or steel-dependent industries. The switch to modern methods has resulted in big increases in efficiency, but at a cost in jobs for steelworkers. At Port Talbot's automated steelworks, for example, output per man is expected to rise from 240 tonnes in 1978 to 300 tonnes in 1985. Meanwhile the total workforce in steel in South Wales will contract by some 10 000 men. (5) From the maps on pages 93 and 96, draw a sketch-map to show the location of each major steelworks by means of a large black dot. Add the coalfield portions of the mining valleys by means of black stripes. Your completed map will show where in South Wales there are special problems of unemployment and redundancy.

For fifty years, ever since the appalling unemployment of the 1930s, great efforts have been made to attract new industries to South Wales. More than 500 factories, producing goods as varied as zip-fasteners, washing machines, ladies' stockings and clocks, have been established by Government action alone. Many of the new works are on industrial estates (*see map*) which give employment to 23 000 people. Other notable developments include large plate-glass and nylon factories in Pontypool. Such factories, using electrical power from the Grid (*p. 264*), have brought new life to many coalfield towns, created new industrial districts, and given a much greater variety of jobs to most parts of industrial South Wales.

Many of the incoming firms are foreign, notably from America and Japan. U.S. examples include the *Alcoa* aluminium works (Swansea), *Ford* engines (Bridgend), *Hoover* washing machines (Merthyr Tydfil), and there are Japanese plastics and T.V. factories in Bridgend and Cardiff. Investment from overseas now

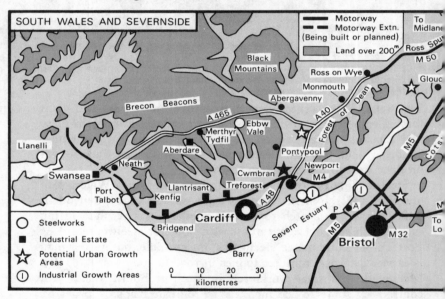

The location of the new industrial estates has been carefully chosen so that the factories are within easy reach of (a) former mining communities in the valleys and (b) redundant steelworkers. From the map below and that on p. 93 (6) say whether the estates are well placed in these respects. Only one of the new estates ((7) *Which?*) is actually within a mining valley. Generally speaking the valleys are too narrow and too isolated for modern factory development. Valley mouth sites, on the other hand, provide flat land with room for expansion and are within easy reach of the South Wales

ports and the region's improved r network. The most important road I is the M4, which makes it possible goods to be exported to the Contin via London Docks—reached in a m 4 hours by heavy vehicles. (8) Wl industrial estates are immediately jacent to the M4? An improved va road from Cardiff to Merthyr Ty favours new factories around Mert and Aberdare. (9) Which other r gives good access to industrial town the head of the valleys and also provi a through route to Swansea?

An oil tanker unloading at Milford Haven, Pembrokeshire.

provides 16% of all jobs in South Wales. Reasons given by foreign industrialists for locating factories in this region include:

government financial help *good communications*
good labour supply *nearness of raw materials*
suitable sites *nearness of component parts.*

Yet another recent development was the siting of a large petro-chemical works near Port Talbot. Milford Haven, a fine ria harbour, has been developed as a modern port able to accommo-date the largest super-tankers. Crude oil is delivered to refineries on the shores of the Haven, or by pipeline to Llandarcy (map, page 268). Oil and petro-chemicals are now the economic mainstay of the whole area from Neath to Pembroke.

In Wales good agricultural land is scarce, yet over a million people live in the mining and manufacturing district of the South Wales coalfield. How do they all get fed? As local produce is insufficient, large amounts of foodstuffs are imported to the region from other parts of Britain and from overseas. Cardiff and Newport are well placed to collect, store, prepare and market these commodities. The map on page 93 shows why.

CARDIFF (290 000) grew up as the world's largest exporter of coal. Now the coal trade has gone and the port handles miscel-laneous cargoes, e.g. iron and non-ferrous ores, timber, grain, refined oils, vehicles and fruit. Cardiff's industries are varied: *con-fectionery, rope making, vinegar making, iron and steel foundries, flour mills, furniture, engine wagon works, breweries, cars, enamel ware, jam manufacture, clothing, cigars, paper.* (*10*) From this list pick out (i) industries concerned with food, drink and everyday needs and (ii) metal working and other industries.

As the capital of Wales, Cardiff is a major financial and business centre. It also has a university college, the Welsh National Museum, and famous sporting centres such as the Arms Park.

Although much of Central and North Wales is barren and unproductive the glaciated Snowdon range, with its numerous lakes and heavy rainfall, is one of the few districts in Britain where hydro-electric power can be generated. There are eight h.e.p. stations. The largest of these, at Ffestiniog and Dinorwic, are **pumped storage** plants. They pump water up to high mountain lakes during the night, when surplus energy is available from the electricity 'grid'. The water is then released during the day to generate power when it is most needed. There are also nuclear power stations at Wylfa and Trawsfynydd.

Although largely bare of soil (*Why?*), the mountains of North Wales contain some rocks which are valuable to man. The main products are *slate*—quarried at Ffestiniog, Llanberis and Bethesda; and *basaltic stone* (for road metal) from Penmaenmawr. The demand for slate has slackened in recent years as clay tiles are now commonly used for roofing. (*11*) Can you suggest why basaltic rocks and slate should be found close together here in the mountains of North Wales? (See page 11.)

The photo (*below*) of Llandudno shows some of the beautiful coastal and mountain scenery of North Wales. Thousands of holidaymakers are attracted to it every year, especially from the industrial towns of the Midlands and the North. Some stay at the many hotels scattered throughout the mountain villages and passes, but most go to the coast resorts indicated on the r above. ((12) Can you name them?) sharp contrast to rugged Snowdonia are low-lying plains of Anglesey. This isla composed of much eroded ancient rocks separated from the mainland by a small valley at the Menai Straits. Holyhea a ferry port for Ireland. (Map, page 2.

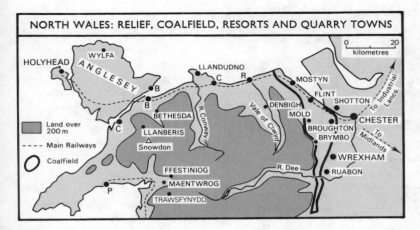

NORTH WALES: RELIEF, COALFIELD, RESORTS AND QUARRY TOWNS

The largest groups of people in Wales outside the South Wales industrial region live in Clwyd, where coal-bearing rocks outcrop on the flanks of the Welsh mountains. Only two collieries now remain open but there are various other well-established industries.

In *Wrexham*, the main industrial centre, old colliery sites have been reclaimed and developed as industrial estates. The town has a remarkable range of modern industries. Examples include fibreglass, synthetic rubber, celanese, chemicals and electronics. Similar electrical and engineering industries in the remote Vale of Clwyd originated as munitions factories. *Brymbo* has a mini-steel mill and there is a large aerospace complex at Broughton.

Deeside, the coastland adjacent to the Dee estuary, has a large aluminised sheet-steelworks at *Shotton*, sheet-iron in *Mostyn*, and chemicals and paper in *Flint*. At nearby *Rhyl* there are factories making garments, electronic equipment and kitchen furniture.

It is suggested that the upper Dee estuary be dammed, to impound fresh water for use in North Wales and Merseyside. The dam would carry a motorway, so enabling more people from North-west England to use North Wales as a dormitory or weekend playground. The reservoir could be used for boating, sailing and swimming.

(*13*) Use the information in this table to make a *large* labelled sketch-map of Industrial North-east Wales. (*14*) Is this region well placed for future industrial expansion? (*Hint:* E.E.C.)

Llanberis slate quarries, near Snowdon.

The Welsh Border Country

Map showing: TO NORTH WALES, TO BALA & BARMOUTH, LAND OVER 200 m, SHREWSBURY, TO ABERYSTWYTH, R. Severn, TO MIDLAND PLAIN, Ludlow, KIDDERMINSTER, TO ABERYSTWYTH, HEREFORD, R. Wye, TO BRECON & CARMARTHEN, TO SOUTH WALES, 0 20 kilometres

Castles, like this one at Ludlow, are numerous in the Welsh border country. They once guarded the routes to and from Wales via such valleys as the Upper Severn and Wye.

In these valleys, and on the plains of Shrewsbury and Hereford into which they open, the people are mostly farmers. Much of the land is permanent pasture, sheep being reared on the higher slopes of the Welsh Uplands and in the Shropshire Hills, and cattle on the lower land. There are particularly rich soils on the Old Red Sandstone of the Hereford Plain. Hereford beef cattle and cross-bred sheep are world-famous.

The Hereford Plain is in the ' rain shadow' of the Welsh Uplands. (*15*) From the figures below draw a climate chart for Hereford and compare it with that for Haverfordwest on page 91. Note the contrasts in the annual temperature range and the total annual rainfall of these two places. The mild conditions in the Plain, with its warm sunny summers and moderate rainfall, favour fruit and vegetable growing. Large quantities of apples, blackcurrants, strawberries, peas and carrots are produced, and cider is made locally. Some cereals and root-crops are also grown, and there are sugarbeet factories at Kidderminster.

The chief market towns of the borderland are Shrewsbury and Hereford. Details of the defensive site of the former, within a meander of the River Severn, are shown opposite.

HEREFORD: Mean Monthly Temperature and Rainfall

	Jan.	Feb.	Mar.	Apl.	May	June	July	Aug.	Sept.	Oct.	Nov.	De‹
°C.	3.9	4.1	5.9	8.3	11.2	14.4	16.2	15.7	13.3	9.7	6.2	4.4
mm	66	46	46	46	56	43	58	60	58	69	69	58

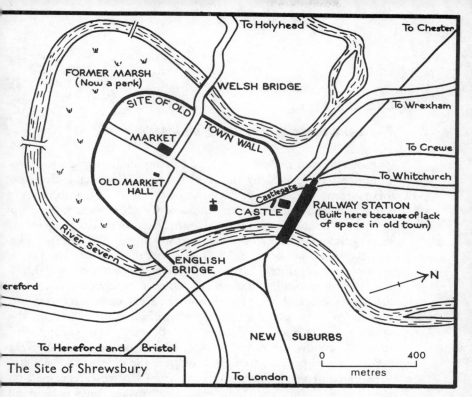

To Holyhead

To Chester

FORMER MARSH
(Now a park)

WELSH BRIDGE

SITE OF OLD

TOWN WALL

To Wrexham

MARKET

To Crewe

OLD MARKET
HALL

To Whitchurch

Castlegate

CASTLE

RAILWAY STATION
(Built here because of lack
of space in old town)

River Severn

ENGLISH
BRIDGE

N

ereford

NEW SUBURBS

0   400

To Hereford and Bristol

metres

The Site of Shrewsbury

To London

Suggest two points where you would
ct to find a steep slope just outside
ld town walls. (Hint: see p. 24) (17)

Describe in a few words the site of the
castle, and suggest why this site was
chosen.

CHAPTER 6

The Lake District and the Pennines

FROM THE map opposite, (*1*) draw a simplified diagram to show the pattern formed by the chief rivers of Cumbria. Describe this pattern. What shape are most of the lakes in Cumbria?

Early in geological time the rocks of Cumbria were heaved up to form a dome, from which streams have been draining in all directions for millions of years, cutting their valleys deep down into the 'cap' of the dome. The pattern they have formed is an example of **radial drainage.** The lower diagram opposite shows how the newer sedimentary rocks which once covered the dome have been stripped away by erosion to reveal ancient igneous and metamorphic rocks. As the rivers which had developed on the sedimentary surface cut deeper and deeper, their valleys continued to follow the radial pattern, and in many cases cut right across the complicated folds and faults of the older rocks beneath.

Millions of years later, in glacial times, an ice-cap formed on the central Cumbrian Mountains. Ice pushed outwards in all directions towards the lowlands, so that the river valleys were gouged and over-deepened by glaciers (see also page 33). When the ice melted, lakes collected in the hollows left on the valley floors. In some cases (like Lake Windermere) water was trapped behind natural dams of morainic debris (page 35).

(*2*) Copy the diagram below into your books, adding the following labels in their correct places:—

GLACIER	TERMINAL MORAINE
ROCK FRAGMENTS WITHIN ICE	OVER-DEEPENED VALLEY FLOOR

Add arrows to show the direction in which the ice is moving. (*3*) Explain the origin of the rock fragments within the ice and describe what part they play in the erosion of the valley floor.

(*4*) Draw a similar diagram showing the depth to which water will collect in this valley after the melting of the ice.

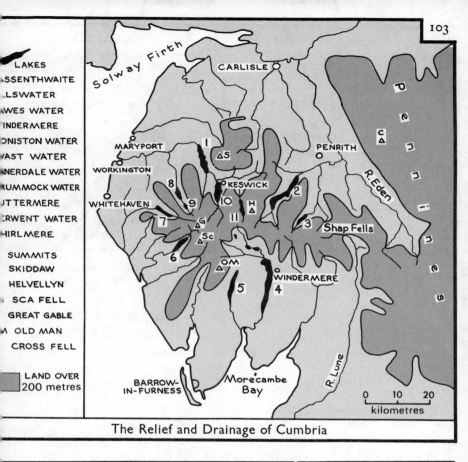

LAKES
BASSENTHWAITE
ULLSWATER
HAWES WATER
WINDERMERE
CONISTON WATER
WAST WATER
ENNERDALE WATER
CRUMMOCK WATER
BUTTERMERE
DERWENT WATER
THIRLMERE

SUMMITS
SKIDDAW
HELVELLYN
SCA FELL
GREAT GABLE
OLD MAN
CROSS FELL

LAND OVER 200 metres

Solway Firth
CARLISLE
Pennines
MARYPORT
WORKINGTON
WHITEHAVEN
KESWICK
PENRITH
R. Eden
Shap Fells
WINDERMERE
BARROW-IN-FURNESS
Morecambe Bay
R. Lune

0   10   20
kilometres

The Relief and Drainage of Cumbria

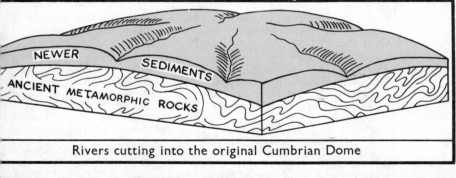

NEWER SEDIMENTS
ANCIENT METAMORPHIC ROCKS

Rivers cutting into the original Cumbrian Dome

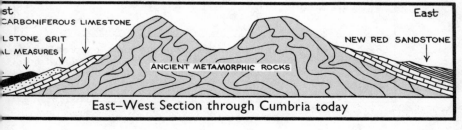

West
CARBONIFEROUS LIMESTONE
MILLSTONE GRIT
COAL MEASURES
ANCIENT METAMORPHIC ROCKS
East
NEW RED SANDSTONE

East–West Section through Cumbria today

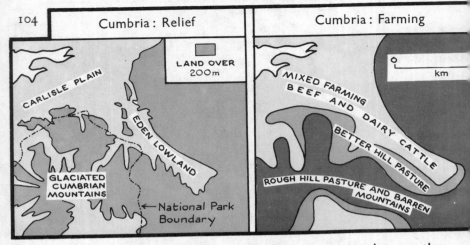

These maps and notes show how the many contrasts between the highland and lowland parts of Cumbria are reflected in the types of farming in these two districts.

FARMING IN CUMBRIA

The Cumbrian Mountains	The Eden and Carlisle Lowlands
Glaciated highland	Glaciated lowland
Igneous, metamorphic and volcanic rocks	New Red Sandstone and boulder clay
Scanty and infertile soils	Fertile soils
Heavy rainfall from Westerlies (over 3750 mm in places) Cold winters	Moderate rainfall—in rain-shadow (750 to 1000 mm) Mild winters
On the 'fells' (barren hills like those in the photograph opposite) SHEEP graze on the rough herbage of better-drained slopes. Arable farming is confined largely to the moist meadows of alluvial valley bottoms. Here grass, clover and barley are the main crops, and some dairy cattle are reared.	MIXED FARMING predominates here. Barley, turnips, swedes and potatoes are grown, and CATTLE REARING is important. DAIRYING is widespread throughout the lowlands and much milk is sent daily to the industrial centres of the north of England, e.g. Liverpool and Newcastle.

Shepherds bring their flocks down from the more exposed, often snow-clad, 'fells' during the winter to graze on the sheltered lowlands. (5) Where else in Britain have we found a similar movement of animals?

A typical Lake District scene. Originally e large lake stretched along that part of e glacial valley floor now occupied by the o lakes—Buttermere (B) and Crummock ater (C). They are divided by a 'delta' (D), built of alluvium dumped in the former lake by mountain torrents. Derwent Water and Bassenthwaite are separated in a similar manner.

The variety of mountain and lake scenery attracts many holiday-makers and climbers to the Lake District. The main centres of the tourist industry—which is one of the chief sources of livelihood for the people—are Keswick, Ambleside and Windermere.

To cater for the flood of summer visitors, camping and picnic sites, car parks, forest walks and nature trails, a wildlife centre, a deer museum and tree-top watch towers have been set up in the Lake District National Park. (*See map.*) (6) Should the number of visitors to such Parks be limited? (7) If so, why?

As the rugged mountains make the building of roads and railways difficult, much of the grandest Lakeland scenery can be reached only on foot.

In great contrast, routes from all over Britain converge on Carlisle. In Roman times a fortress there guarded the western end of Hadrian's Wall. Today it is a great route centre. Its industries include engineering, tyres and food processing (notably biscuits, sausages and pork pies).

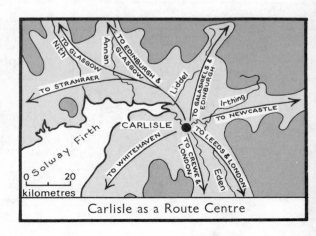

Carlisle as a Route Centre

Cumberland:
Coalfield and Iron Mines

MARYPORT
COCKERMOUTH
WORKINGTON
DISTINGTON
KESWICK
WHITEHAVEN
×CLEATOR
A 66(T)
×EGREMONT
To M6 →
×BECKERMET

MILLOM
×
⊗ COALFIELD
×
× IRON ORE MINES
(Now mostly worked
out)
BARROW-IN-FURNESS
×DALTON
Furness

0   20
Kilometres

Between the Cumbrian Mountains and the Irish Sea, scenes of Lakeland beauty are replaced by the scars of industry—coal-tips, steel mills, factories and shipyard cranes. Industry first developed here due to the presence of coal and high-grade iron-ore deposits (see map). Now only one colliery remains and iron-ore production is confined to the haematite mines south of Egremont. The small iron and steelworks at Barrow-in-Furness and Workington now depend on imported ore, pig-iron, scrap metal and coke. A third steelworks at Millom closed in 1970 because of poor transport facilities thus highlighting the main problem of west Cumbria—remoteness. Improvements to the A66 road linking Workington to the M6 motorway have lessened the problem. This road especially favours the new industrial estates near Workington. Details of industries in the principal towns are given opposite.

In recent decades several new industries have taken root in Cumbria, and many new firms have arrived since the region was designated a Special Development Area in 1966. ((8) *Why? What advantages does such designation give?*) These new industries have largely superseded the region's old economic base of coal and steel: details are given opposite.

Cumbria has become famous as the home of the world's first nuclear power station—the Calder Hall plant near Whitehaven. On an adjacent site the Windscale factory provides nuclear fuel for all of Britain's and some overseas nuclear power stations.

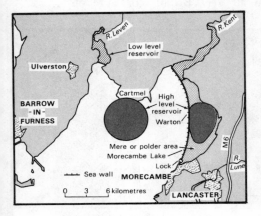

BARROW
-IN-
FURNESS
Ulverston
R. Leven
Low level
reservoir
R. Kent
Cartmel
High
level
reservoir
Warton
Mere or polder area
Morecambe Lake
Lock
Sea wall
MORECAMBE
M6
R.
Lune
0   3   6 kilometres
LANCASTER

Windscale is also the U.K. centre for treating spent fuel, from which some uranium and plutonium are recovered. Over 7000 people are employed at Calder Hall and Windscale.

Cumbria has immense quantities surplus fresh water. A suggested sche (*see map*) would trap and store suffici water from the rivers Leven and Kent meet all foreseeable domestic and dustrial demands throughout northe England.

Merchant Ships: Total Tonnage Launched* 1973–78	
Belfast . .	1 750 305
Clyde . . .	1 413 880
Humber . .	60 885
Mersey . .	310 701
Tees . . .	453 466
Tyne . . .	1 199 200
Wear . . .	1 500 756

Lloyd's Register

In these notes the leading British centres of *merchant* ship-building are placed in alphabetical order. (9) Rewrite the list in order of tonnage launched.

(10) Draw a map of Britain and mark on it all of these ship-building districts, with a number after each name showing its position in the list you have made.

MAIN INDUSTRIAL TOWNS OF CUMBRIA

BARROW-IN-FURNESS's principal industries are *shipbuilding and engineering.* The giant Vickers concern, which employs 14 000, has been building ships since 1873. The world's first nuclear-powered submarine and Britain's first Polaris submarine were built here, and the shipyard now specialises in high-technology warships. Barrow's steelworks and harbour were originally constructed to use and export Furness iron ore. Today all steel-making raw materials are imported, and the Vickers works produces steel by an electric arc furnace. A wide variety of armaments is made, as well as oil-drilling rigs and other 'off-shore' equipment. Barrow also has Europe's largest *tissue-making* plant. (11) Using what imports?

WORKINGTON has old-established *iron and steel* and *engineering* industries. Steel-making ceased in 1974, but iron is still made, there is a rail-rolling mill, and the nearby Distington engineering works produces alloy steels and steelworks equipment. Engineering firms in Workington also make railway track and aircraft components and hydraulic machinery, and the Leyland National bus company produces 2000 buses p.a.

The wide range of newer products from Workington's industrial estates include *packaged foods, metal containers* and *cigarette filters.* A paper and board mill uses timber from Forestry Commission plantations in N. England and S. Scotland, and the plentiful local supply of pure water from the R. Derwent. Workington is the only sizeable commercial port between Liverpool and Glasgow.

WHITEHAVEN has important *chemical* works, making goods such as detergents, shampoos, waxes and cement. Anhydrite (gypsum) for this industry is mined at Cocklakes, near Carlisle. Other raw materials, e.g. phosphate rock, are imported. Other diverse employment in Whitehaven is provided by the manufacture of *foodstuffs* (Rowntrees and Quaker Oats), *carpets, clothing, paper* (Thames Board) and *pharmaceuticals.*

Lesser industrial centres include KENDAL and COCKERMOUTH (*leather and footwear*) and MARYPORT (*engineering*). For CARLISLE see p. 105.

(12) Use these notes to make a large, labelled sketch-map. Give your map an appropriate title.

108

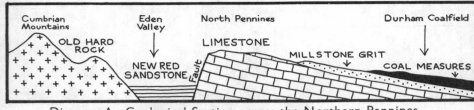

Diagram A : Geological Section across the Northern Pennines

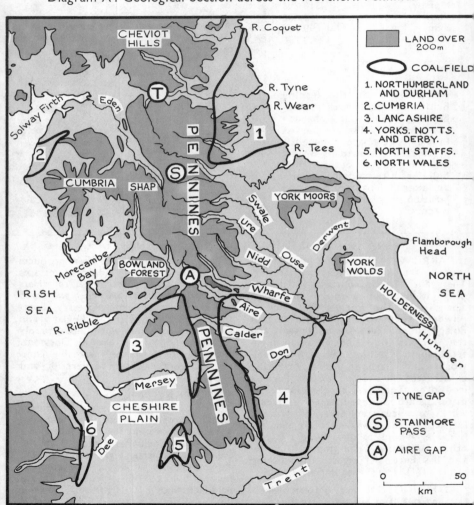

Physical Map of Northern England

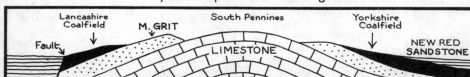

Diagram B : Geological Section across the Southern Pennines

For 260 kilometres—from the valley of the Tyne to the Trent Valley in Derbyshire—the Pennine Uplands form the ' backbone ' of England. Varying in width from 30 to 60 kilometres and in places rising well over 700 metres, these Uplands form a barrier between the lowlands of Lancashire and Yorkshire—breached in places by railway tunnels and by three important gaps:

1. Tyne Gap, between Carlisle and Newcastle.
2. Stainmore Gap, linking the upper Eden and Tees Valleys.
3. Aire Gap, between Preston and Leeds, and linking the industrial districts of Lancashire and West Yorkshire.

(*13*) From the map opposite, which of these gaps appears to be the most difficult ?

The rocks of the Pennines are Carboniferous Limestone, Millstone Grit and Coal Measures. As the diagrams show, these have been upfolded and eroded. All of the coal-bearing rocks and much of the Millstone Grit have been removed from the top of the anticline (upfold: see page 13), and as a result the coal-bearing strata outcrop along the flanks of the Pennines. The National Coal Board classifies the coalfields shown opposite as follows:— (a) Western Area (Cumbria, N. Wales, Lancs. and Staffs.); (b) Northumberland and Durham; (c) N. Yorks., Barnsley Area, Doncaster Area, S. Yorks., N. Derbyshire, N. Notts. and S. Notts. For geographical purposes the names used on the map are more satisfactory. (*14*) Why?

The general appearance of the Pennines is of a flat-topped **plateau** (uplifted plain). The sketch above is of Pen-y-Ghent, a higher part of the Central Uplands. (*15*) Make a copy of it and show where you think is the junction between Millstone Grit and Carboniferous Limestone. (The diagram below will help.)

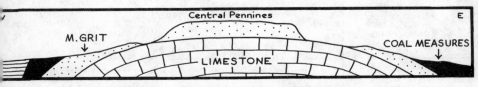

Diagram C: Geological Section across the Central Pennines

Limestone 'pavement' above Malham Cove, Yorkshire.

Thick beds of almost pure limestone lie at the ground surface in the northern and southern parts of the Pennines. Here are

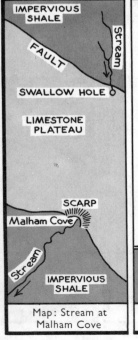

IMPERVIOUS SHALE

FAULT

Stream

SWALLOW HOLE

LIMESTONE PLATEAU

SCARP

Malham Cove

Stream

IMPERVIOUS SHALE

Map: Stream at Malham Cove

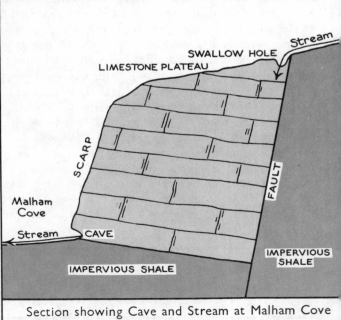

Stream

SWALLOW HOLE

LIMESTONE PLATEAU

SCARP

FAULT

Malham Cove

Stream

CAVE

IMPERVIOUS SHALE

IMPERVIOUS SHALE

Section showing Cave and Stream at Malham Cove

found many of the features of 'karst' countryside described on pages 26 and 27 (which should now be read again). Notice in the photograph: *1. The scanty vegetation and many barren rock surfaces. 2. The steep escarpment.*

A small stream (a tributary of the River Aire) flows from the foot of this escarpment, and its course to this point is charted in the diagram and map opposite. (*16*) Look carefully at these and then describe the path taken by the river, from its source until it emerges from within the limestone. Include the following words in your account: *impervious; surface stream; fault; limestone; joints; porous; swallow-holes; underground stream.*

Most of the highland in the Pennines is too damp, cool and infertile for cultivation, but in the limestone districts there is a thin covering of turf which makes good pasture for sheep.

In the Central Pennines, where the so-called 'Millstone Grit' series of rocks lie at the surface, drainage is poor. This is because layers of impervious shale occur at intervals between the main beds of sandstone. There are many peat-bogs, and the poor vegetation is of so little value that the district is practically uninhabited—although *not* unimportant. As Millstone Grit is insoluble, the water flowing from these bleak moors is exceptionally 'soft' and pure. Unlike the 'hard' waters from the limestone districts, the streams from the Millstone Grit could be used for washing wool fleeces and cloth. This was one important reason why large numbers of early woollen mills were built in valleys along the flanks of the Central Pennines. ((*17*) *To what other vital use were the streams put, before the Industrial Revolution?*)

Today millions of litres of pure water are stored in reservoirs on the Millstone Grit uplands, for factory and home use in the near-by industrial areas of Lancashire and Yorkshire. Below is a sketch of Ryburn Reservoir, which supplies the city of Wakefield.

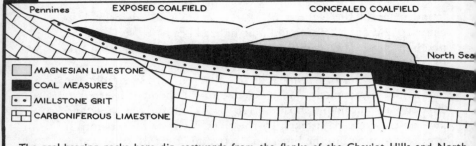

Geological Section across the Southern Part of the Durham Coalfield

Pennines   EXPOSED COALFIELD   CONCEALED COALFIELD

North Sea

☐ MAGNESIAN LIMESTONE
■ COAL MEASURES
∷ MILLSTONE GRIT
⊞ CARBONIFEROUS LIMESTONE

The coal bearing rocks here dip eastwards from the flanks of the Cheviot Hills and North Pennines.   To the west, where the Coal Measures outcrop at the ground surface, is the **exposed coalfield**; to the south-east lies the **concealed** field, where the coal dips beneath Magnesian Limestone and other rocks.

CHAPTER 7

Industrial North-East England

For CENTURIES the fortunes of this region have been linked with coalmining.   During the 19th and early 20th centuries an enormous coal export trade grew up, favoured by the coalfield's location on the coast.   Today coal exports have dwindled and the coalmining industry has shrunk, involving a rash of pit closures. During the 1960s these closures mainly affected the older mines in west Durham. ((*1*) Why were these mines *older?* Hint— depth of coal seams.) Now many newer mines to the east have closed, and future coal production will come chiefly from seven large modernized pits on the coast.   These pits are best located to reach the estimated 600m. tonnes of high-grade coal which still lies deep underground in the North-east, much of it below the bed of the North Sea.   From 1978 the region's re-organized mines will yield about 16m. tonnes of coal annually, and so they could enjoy a working life of at least (*2*) how many years?

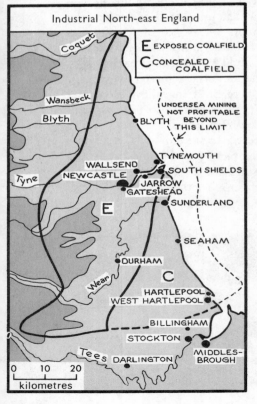

Industrial North-east England

E EXPOSED COALFIELD
C CONCEALED COALFIELD

Coquet

Wansbeck

Blyth

Tyne

UNDERSEA MINING
NOT PROFITABLE
BEYOND
THIS LIMIT

BLYTH

TYNEMOUTH
WALLSEND   SOUTH SHIELDS
NEWCASTLE
JARROW
GATESHEAD
SUNDERLAND

E

SEAHAM

DURHAM

Wear

C

HARTLEPOOL
WEST HARTLEPOOL

BILLINGHAM
STOCKTON

Tees   DARLINGTON

MIDDLES-
BROUGH

0   10   20
kilometres

From the table on the right (3) draw bar diagrams to illustrate the contraction in (a) coal production, (b) colliery employment and (c) coal exports between 1961 and 1978. Not enough pits remain open to absorb all of the

Northumberland & Durham Coalfield			
	1961	1971	1978
Coal Output . (million tonnes)	33.2	15.5	16.2
Total Colliery Employees .	116 000	48 000	36 000
Total Coal Exports (million tonnes)	0.84	0.13	0.61

Despite a huge reduction in manpower, coal-mining in the North-east is still important. Output has picked up since the oil crisis of 1973–4.

surplus labour and thus the provision of new employment is the region's most urgent problem. It is made difficult by the isolation of many former mining villages. Some displaced miners find new jobs in factories within daily travelling distance of their homes, but others move with their families to new *industrial estates* adjacent to the A1 Motorway. (*See photo.*) Details of these estates are given on p. 117. Partly to attract new firms into the North-east the whole region is being given a ' face-lift ', including the levelling of hundreds of ugly coal-tips. The reclaimed land is used for farming or is converted into public parks and playing fields. Red shale from the tips is used for building roads and embankments.

Washington New Town. *The New Town is made up of 18 relatively self contained units called Villages. Each Village has about 1350 houses catering for 4500 people. There is a Village centre with three to five shops, a meeting hall and a pub or club together with a two form entry Primary School. Ample open space is included in each Village. This picture shows a safe play area for young children at Donwell Village.*

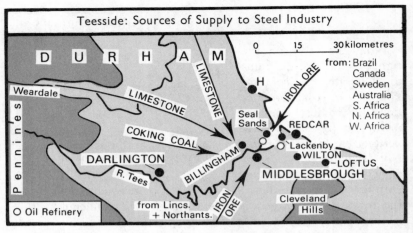

Teesside: Sources of Supply to Steel Industry

IRON ORE from: Brazil, Canada, Sweden, Australia, S. Africa, N. Africa, W. Africa

Teesside (400 000) is one of the leading iron and steel producing districts in Britain. The early foundries obtained nearby all the raw materials needed for iron smelting, using (i) iron ore from the Cleveland Hills, (ii) best-quality coking coal from south-west Durham and (iii) limestone from Weardale and east Durham. In time, as the Cleveland supplies were depleted, it became necessary to import iron ore. Such imports were made easy by the coastal position of the iron and steel mills. Similarly, heavy and bulky products like steel girders and engines, which are so typical of this district, could be shipped abroad without a costly overland journey. The Tees estuary, with its wide, deep-water channel flanked by extensive mud-flats where factories can be built, is likely to be a major growth point for heavy industry. (4) What added locational advantage does this estuary have now that Britain has joined the Common Market?

One-third of all British steel will eventually be made at Redcar and Lackenby, where a new works includes the most modern coke ovens, the biggest blast furnace ever built, and an iron-ore terminal able to dock large bulk carriers. Another new mill will make heavy plate for North Sea pipelines. Much Teesside steel is used in the North-east's shipbuilding and engineering industries. New engineering and various light industries are taking root on trading estates near Middlesbrough, Hartlepool and Darlington.

Hartlepool (100 000) has recently lost its shipbuilding and steelmaking industries and its export trade in coal, whilst the iron ore formerly imported there is now handled on Teesside. Port activities now centre on imports of Scandinavian timber and wood pulp, the latter for despatch to paper mills in Lancashire. The town's main industry is engineering: products include pipelines and drilling rigs for the North Sea oil- and gas-fields.

Darlington (85 000) grew up as a railway engineering town, but its famous steam locomotive and wagon workshops are now closed. The town's strategic location on the A1(M) and at the 'gateway' to Teesside has favoured the growth of service industries such as wholesale distribution, road haulage, construction, finance and insurance. The main industry is engineering: products include diesel engines, bearings, containers and pipeline valves.

This ethylene storage sphere and 'cracking' plant typifies Teesside's other main fast-growing industry—chemicals. The industry originated at Billingham, where local deposits of rock salt and gypsum are used to make such products as sodium carbonate, chlorine and sulphuric acid. More recently production has concentrated mainly on petro-chemicals.

Developments in petro-chemicals, coupled with the exploitation of North Sea oil (see text) makes Teesside a dynamic industrial region. Many 'white collar' jobs are provided in the new refineries and chemical works, and there is a steady demand for equipment from local electronics and engineering firms.

The Billingham factories of Imperial Chemical Industries cover more than 500 hectares, and have grown into one of the world's largest plants for the production of heavy chemicals from petroleum and coal: in addition a vast newer works at Wilton makes terylene cloth, plastics and other synthetics. The Billingham and Wilton works are linked under the Tees by pipe-lines which carry petroleum for 'cracking' prior to the manufacture of synthetic chemicals. Seal Sands (*see map*) is the terminus for 'Norpipe Teesside', which brings North Sea oil from the Norwegian Ekofisk field to be processed and transhipped. Refinery facilities at Seal Sands include the ten largest oil storage tanks in Europe. The nearby *Monsanto* textile plant makes *acrylonitrile* for the production of plastics and synthetic fibres.

Other chemicals made on Teesside include oxygen, chromium salts and fertilizers. A nitrate fertilizer industry developed as a branch of petro-chemicals, and now potash is also available from a deep mine near Loftus. These newly-discovered potash beds are very rich: their working, however, is strictly controlled to avoid polluting and disfiguring the beautiful countryside of the North Yorkshire National Park.

A merchant ship ready for launching on Tyneside. All types of vessels are built along this famous estuary, including warships, oil-tankers, bulk cargo carriers, car ferries, liquefied gas carriers and oil drilling platforms.

This is a familiar scene in the North-East, where a tradition of shipbuilding reaches back at least five centuries. Small wooden vessels were built at first, to carry away coal from the coastal workings of the Northumberland and Durham coalfield. Timber for the ships came from Scandinavia, followed in time by high-grade Swedish iron ore. In return coal was sent to Norway and the Baltic countries. In the 19th century, when iron replaced wood in ship construction, the shipyards on Tyneside prospered, for they had the advantage of a local iron industry. Nowadays sup-

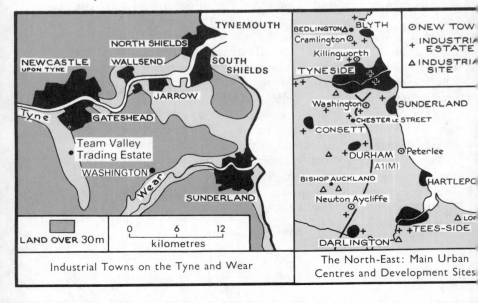

Industrial Towns on the Tyne and Wear

The North-East: Main Urban Centres and Development Sites

plies come mainly from a large steelworks at Consett, using iron ore imported via South Shields. A new mini-mill in Jarrow also imports ore to produce 0.8m tonnes of steel in electric arc furnaces. Tyneside was one of the hardest hit of British industrial districts during the 'depression' years in the 1930's. The chart shows the decline in shipbuilding at that time. Relative prosperity returned in the 1950's but since then the shipbuilding industry has once again fallen on lean times: the reasons are explained on page 81. Chronic unemployment in the shipyards, coupled with the run-down in the coal industry (see page 113), has led to massive Government and local efforts to attract new industries to the North-East. The old-established engineering industry—formerly concerned mainly with shipbuilding and ship repairing—has expanded to include a bewildering range of activities (see below), and engineering looms large among the 200 firms now established on the new trading estates and towns shown opposite. These estates provide a remarkable variety of employment for about 100 000 people. Industries include the manufacture of factory-built houses, processed foods, razor blades, toilet products, cartons and chemicals.

Typical Engineering Products of the North-east	
Switchgear, Cranes	Printing presses
Generating plant	Machine tools
Transformers	Boilers, Engines
Car components	Pumps, Telephones

About 9000 engineering workers in the North-east are benefiting from the exploitation of North Sea oil and gas. Work involves the manufacture and servicing of platforms and modules and the laying and maintenance of pipelines.

In Sunderland (Wearside) shipyards and coalmines still employ 35% of the male work force, but light industrial jobs are also available in the nearby New Town of Washington.

(5) Show the following information as a bar diagram:

Percentage employment in the North-east:—agriculture 2; ship-building and marine engineering 4; iron and steel 4.5; chemicals 4.5; construction industries 7.8; coalmining 8.7; other manufacturing 23; non-manufacturing activities 46.

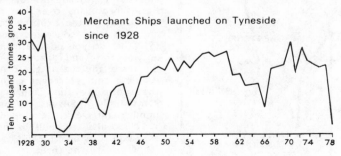

Merchant Ships launched on Tyneside since 1928

CHAPTER 8

The Yorks., Notts. and Derby Industrial Region

THE Yorks., Notts. and Derby Coalfield has for long been one of Britain's most important sources of wealth. It is the country's largest coalfield, but like all others in Britain it has suffered contraction both in output and in manpower. However, so much good quality coal still remains in this field that pit closures and labour redundancies have been less severe here than elsewhere. The present labour force of 100 000 produces almost half of the

The Yorks., Notts. and Derby Coalfield:—is the largest coalfield in Britain; yields high-quality coals of all kinds ((*1*) *name three main types*); dips gently below sandstones towards the east; is little affected by faulting ((*2*) *how does this affect mining?*); has never been fully charted towards the north-east; was first worked along its western margin ((*3*) *why? see diagram, page 124*).

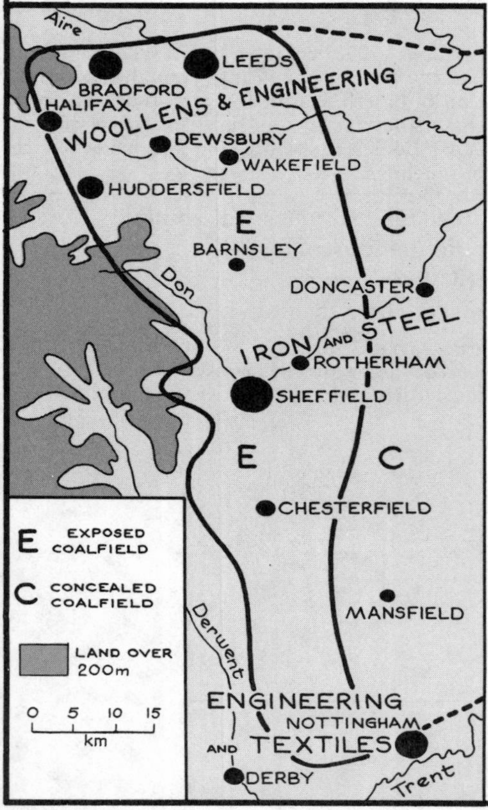

(*4*) Study the maps carefully and work out:

(*a*) the length of the coalfield from north to south; (*b*) the width of the *exposed* coalfield at Barnsley; (*c*) the greatest distance of a colliery from the exposed coalfield and the depth of the coal below ground at this point; (*d*) the number of collieries producing over 1 m. tonnes of coal in (*i*) the exposed coalfield and (*ii*) the concealed coalfield. (*5*) What do you deduce from your answers to (*c*) and (*d*) about:—(*f*) the quality of the coal in the concealed coalfield and (*g*) recent changes in the landscape around Doncaster and Mansfield? (*h*) Suggest reasons why Immingham Dock, developed since 1912, has displaced Hull as the main coal-exporting port for this coalfield. (*6*) What are the three main industries of the coalfield?

The Yorks., Notts. and Derby Coalfield

Aire
LEEDS
BRADFORD
HALIFAX
WOOLLENS & ENGINEERING
DEWSBURY
WAKEFIELD
HUDDERSFIELD
E   C
BARNSLEY
Don
DONCASTER
IRON AND STEEL
ROTHERHAM
SHEFFIELD
E   C
CHESTERFIELD
MANSFIELD
Derwent
ENGINEERING
NOTTINGHAM
AND TEXTILES
DERBY
Trent

E   EXPOSED COALFIELD

C   CONCEALED COALFIELD

LAND OVER 200m

0 5 10 15
km

total output of U.K. coal. Production is being concentrated on certain easily-worked pits which produce the high calorific and coking coals suited to modern industry. Recent test borings have located coal seams up to 4 metres thick to the north of Selby, with reserves of over 600m tonnes. By the mid 1980s a new 'Selby Project' drift mine at Gascoigne Wood will produce 10m tonnes p.a.—i.e. (7) what percentage of all Britain's coal? (*See p. 268.*)

One major local use for this coal is to generate electricity in the important coal-fired power stations indicated in the map below. The giant Ferrybridge station, for example, uses no less than 18 000 tonnes of local coal every *day* and 3 822 000 litres of cooling water every *minute*. The availability of ample supplies of water from the Trent, Aire and Calder, and the existence of good flat building sites in the lower flood plains of these rivers are other reasons why more new power stations have recently been built in this region than anywhere else in Britain.

Main Collieries 1979

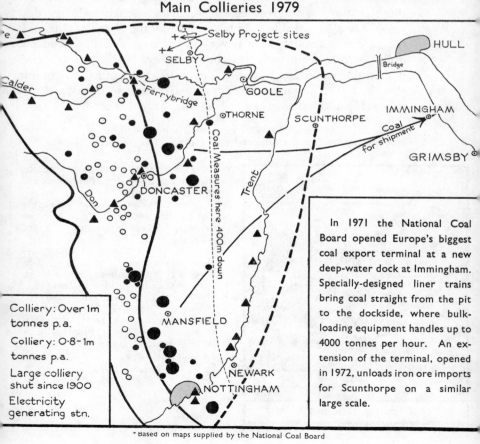

Selby Project sites

SELBY

HULL

Bridge

Calder

Ferrybridge

GOOLE

THORNE

SCUNTHORPE

IMMINGHAM

Coal for shipment

GRIMSBY

Coal Measures here 400m down

Don

DONCASTER

Trent

MANSFIELD

NEWARK

NOTTINGHAM

Colliery: Over 1m tonnes p.a.

Colliery: 0·8-1m tonnes p.a.

Large colliery shut since 1900

Electricity generating stn.

In 1971 the National Coal Board opened Europe's biggest coal export terminal at a new deep-water dock at Immingham. Specially-designed liner trains bring coal straight from the pit to the dockside, where bulk-loading equipment handles up to 4000 tonnes per hour. An extension of the terminal, opened in 1972, unloads iron ore imports for Scunthorpe on a similar large scale.

Based on maps supplied by the National Coal Board

WEST YORKSHIRE is by far the most important woollen manufacturing region in the country. (8) What percentage of Britain's total consumption of raw wool is spun there? This dominance has been achieved since the late 18th century. Before that time wool-

Amount of Clean, Raw Wool consumed at the spinning stage (000 tonnes)	
West Yorkshire . . .	164.6
Scotland (mostly Tweed Valley)	23·8
West of England & South East	6·9
Other	105·0
Total . . .	300·3

len cloth was made on simple hand-operated machines in all parts of Britain where sheep were reared and where soft, lime-free water was available for washing wool and dyeing cloth. With the development of mechanised factory production, output became concentrated first in places with swift-flowing streams capable of turning cumbersome water wheels, and later in places where supplies of coal were near at hand. Many districts once famous for their woollens fell into decline.

West Yorkshire, by contrast, prospered greatly as a result of these changes and for several reasons its woollen industry developed to a remarkable extent. In parts of the North of England the medieval guilds* were never very strong, and new woollen mills could be built with little interference. The early weavers, too, could obtain an abundant supply of wool from the Pennine sheep. Other important facts are apparent from the map: study it closely and then answer the questions printed alongside it.

* Guilds were associations of craftsmen which sought to regulate the amount and quality of cloth produced and fix a fair price for its sale.

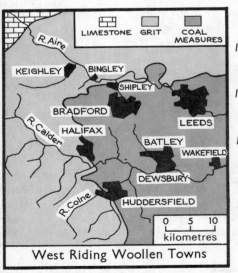

West Riding Woollen Towns

9. Why do most of the streams draining this part of the central Pennines contain soft water?
10. Do you consider this was a suitable district for cloth making in the age of water-driven machinery? If so, why?
11. Why are the majority of the industrial towns today confined to the *lower* Pennine slopes between the Aire and Colne-Calder valleys?
12. Why are there so few industrial towns on the River Aire, whereas they are numerous on its tributary streams, e.g. Keighley, Bingley, Shipley. (Hint: trace the Aire towards its source.)

(Above) A Pakistani worker in a Bradford woollen mill. (Below) Halifax, West Yorkshire. Note the woollen mills with their tall chimneys; the rows of older terrace houses; the newer suburban housing estates and the blocks of high-rise flats.

IMPORTANT TOWNS OF WEST YORKSHIRE

LEEDS (496 000). Largest town and 'capital' of the region. A *textile* centre since 1600. Contains about 200 factory-made *clothing* firms, e.g. Hepworths and Burtons. Large clothing sales to Common Market countries.

Much urban renewal, especially in Central Business District, where shops and offices provide 150 000 jobs.

Good road and rail links (p. 281) make Leeds a fine *distribution and service centre:* hence many mail order houses, banks, insurance brokers, accountants, marketing agencies and so on.

New industrial estates. Over 1800 diverse industrial activities, but mainly *engineering, chemicals* and *printing.*

Leeds University the main centre of woollen textile research.

BRADFORD (294 000). On small south bank tributary of R. Aire. Once the 'capital of wool', now has only 25% of work force in *textiles.* Famous Wool Exchange: still the chief marketing centre for wool and cloth. Manufacture of *worsteds* and *woollens.* Many spinning, weaving, dyeing and finishing mills.

Electrical, textile and general *engineering.*

Recent growth of *distribution and service activities,* e.g. Inland Revenue centre at Shipley.

HUDDERSFIELD (131 000). Grew up at convergence of several valleys tributary to the R. Calder. (*(13)* What was advantage of this position?)

Manufacture of: *woollens* and world's finest *worsted cloth; carpets; dyestuffs.* Work in *pattern designing.*

Textile machinery and other *engineering,* e.g. power station equipment and vehicles.

Chemicals (large I.C.I. factory).

HALIFAX (91 000), has lost 14 000 textile jobs in past two decades. Only 5% of town's workers now in this industry. However, *heavy woollens,* e.g. *carpets* still important.

Textile machinery. Rapid growth of general *engineering,* especially *machine tools.* Also *food processing* (e.g. beer and confectionery) and *furniture.*

DEWSBURY (51 000), BATLEY (42 000) and **WAKEFIELD (59 000).** Chief towns for manufacture of 'mungo' and 'shoddy' —made from waste cloth and knitted goods. Dewsbury and Batley important for *heavy woollens,* e.g. blankets and thick overcoats.

Wakefield, an old woollen town, now has great variety of industries, including: *engineering, electronics* and *food processing* (e.g. Frigoscandia) and *warehousing.* The town is well placed to gain from development of the nearby Selby Coalfield (p. 118).

West Yorkshire: Districts & Motorways

For nearly two centuries the woollen industry, and to a lesser extent coalmining, provided the economic mainstay of West Yorkshire. In recent decades both industries have suffered serious contraction. The woollen industry, for example, has lost nearly 50% of its jobs since 1970 (*see text*). Even so there are still some 400 woollen firms in these towns employing over 50 000 people (only engineering employs more), and one in five of the workforce is in woollens, clothing or coalmining. Despite financial help from the government, the textile industry seems likely to decline still more, but coalmining is expanding again, and Knottingley has one of Britain's largest and most modern pits.

In recent decades the Yorkshire woollen industry has not been able to retain its share in world markets. Too many small firms, using old-fashioned methods and machinery, have been trying to sell the same products in competition with each other and with better equipped foreign producers. In an attempt to bolster output the Yorkshire mills gave jobs to 30 000 Indian and Pakistani immigrants who, unlike many local workers, were willing to work night shifts. Round-the-clock production has improved efficiency but even so many smaller firms have closed or have been taken over by larger companies. The bigger firms can afford to install new labour-saving machinery so that, for example, one man can now operate 20 looms whereas formerly each loom had its attendant. Employment in textiles has thus contracted from 50% to 12% of all jobs in West Yorkshire. The woollen industry is also adapting to changes in fashion by making more lightweight worsteds and mixtures of wool and man-made fibres.

Great efforts have been made to attract new industries into West Yorkshire to offset the contraction in woollens. The towns themselves are becoming more pleasant places, for new clean-air laws forbid factory owners and householders from emitting black smoke from their chimneys: Bradford, for example, became a smoke-free city in 1973. Based on a long tradition of textile machinery manufacture throughout the region, many firms have branched out into other types of engineering. In Bradford no less than 240 of the city's 1400 firms make engineering products ranging from nuts and bolts to radio and television equipment. Other growing industries include mail order firms and tufted carpets. Trade and industry has been stimulated by the new M62 motorway route across the Pennines. (Map p. 281.)

Oil-free compressors being assembled in a Bradford engineering works.

The skill with which this *SHEFFIELD* steel worker is grinding an aircraft casting is part of an inheritance of craftsmanship reaching back over 800 years. Iron was smelted in the Don Valley as early as the 12th century and Sheffield has been famous for its high grade-steel for over 400 years.

Raw materials for the early Don Valley iron industry were available locally.

Raw Material	Source and Comments
Iron ore	'black-band' variety in the Coal Measures near ground surface and easily quarried
Charcoal	abundant timber from forests in Don Valley
Limestone. Gannister Millstone Grit	flux for the blast furnaces hard sandstone for lining furnaces ⎫ from Pennines for millstones to sharpen cutlery ⎭

At first the ironworks depended upon water power:

(*a*) to work bellows which forced an air draught through the furnaces; (*b*) to drive hammers which beat out the iron; (*c*) to turn grindstones for sharpening cutlery.

(*14*) From the map opposite suggest why many early iron-mills were built just west of the present site of Sheffield.

Later, when coke instead of charcoal was used for smelting, and steam replaced water power in the mills, the Sheffield area was fortunate in having near-by supplies of good-quality coking and heating coals.

Although local 'black-band' iron supplies were used up by the 16th century, Sheffield steel was of such good quality that demand for it grew steadily. Using imported (Spanish and Swedish) iron ore, steel mills multiplied in the Don Valley between Sheffield and Doncaster. The main facts about present-day iron and steel

Central Pennines

NEW

INDUSTRY IN THE DON VALLEY

Sheffield district: *light steel goods*, especially cutlery and cutting instruments (for butchers, bakers, painters, glaziers, plumbers etc.); fine *steel forgings* and high quality *alloy steels;* high-grade *engineering products*, e.g. hand engineers' tools. Some very *heavy fine steels*, e.g. rotors for power stations.

Sheffield's steelmaking capacity, which rose from 2.4m tonnes to 3.4m tonnes in the 1970s, is being further increased by the production of more stainless steels. However, the enlarged and modernised steelworks employ 50 000 fewer men. The city's other industries are therefore vital, for Sheffield is dangerously reliant on steel and metal manufacturing. For example, stainless steel—though famous —faces keen competition from aluminium and plastics. Alternative employment includes food processing (Batchelor's foods and meat spreads), and the making of graphite (for electric arc furnaces) and transport containers. Growing indus-

trial estates, e.g. at Mosbrough and Allencliffe, are absorbing many skilled ex-steelworkers.

Rotherham to *Doncaster:* heavy steel goods; castings for turbine engines, armour plating, steel plate for ships, etc. New bar mill at Rotherham. Large railway workshops.

The railway plant now makes general as well as railway engineering equipment. This diversification is typical, especially of the district around Doncaster where there is a growing range of light industries including:—wire, ropes, agricultural implements, glassware, plastics and knitwear. Doncaster's location adjacent to the M1 and M62 motorway intersection favours commercial growth.

Coalmining in the Sheffield-Rotherham-Doncaster region gives employment to 37 000 men. Millions of pounds are being invested—the ten collieries around Doncaster alone have a minimum reserve of 440m tonnes.

production in the Don Valley are given above. Chesterfield (map, page 118) also has iron and steel works and makes mining machinery, furnaces and tubes.

The diagram below shows a section from the central Pennines to the Humber estuary, approximately along the line of the Don Valley. (*15*) Make a *large* copy of it and add the following:—

1. Complete the labelling of the strata.
2. Three industrial towns are shown:— (*i*) Doncaster (on concealed coalfield); (*ii*) Sheffield (on exposed coalfield); (*iii*) Scunthorpe— an iron-mining town on the nearby Jurassic escarpment. (*See also Ch. 12*). Label these towns.
3. Place a letter *E* on your diagram to show approximately where the early iron-mills which used water power were located.
4. The Don Valley steel industry may be divided into 'light' and 'heavy' sections. On your diagram print LIGHT and HEAVY over the appropriate districts.
5. Using arrows indicate and label *one* dip slope and *one* scarp slope.

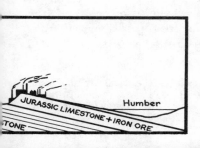

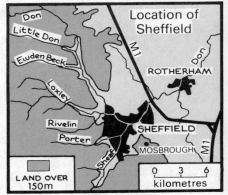

NOTTINGHAM and *DERBY*. Nylon stockings and Rolls Royce engines, railway locomotives and elastic: these are a few of the great variety of products manufactured in the Nottingham-Derby district. (*16*) Look carefully through the fuller list which follows and re-

Railway engineeri... workshop, Derby.

arrange the articles under three headings—TEXTILES, ENGINEERING, OTHER INDUSTRIES. You will then see that these miscellaneous goods are not quite so unrelated as one might imagine.

stockings, tobacco, rayon cloth, socks, cycles, locomotives, lace, tapes, drugs, boots and shoes, aero engines, steel rails, jumpers, textile machinery, braid, shoe laces, elastic, railway carriages, terylene cloth, pottery, nylon, paint, bleaching, typewriters, bathing costumes, curtain material, underwear, scarves, dynamos, pumps, power station boilers, steel cord, switchgear, desalination plant, lawn-mowers.

Specialised textile trades, such as hosiery and lace manufacture, grew up here largely due to inventions by local craftsmen; e.g. in

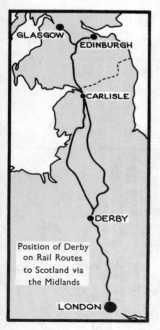

Position of Derby on Rail Routes to Scotland via the Midlands

1589 William Lee, a Nottingham parson, invented a machine for knitting stockings; later came Strutt's machine in Derby for making ribbed hose and Heathcote's machine for weaving plain lace net. In 1843 Jacquard invented a loom for weaving fabrics into elaborate patterns and soon the Nottingham-Derby district became famous for its high-quality hosiery and knitwear. (*17*) Textile manufacture in Nottingham and Derby is concentrated largely on making cotton ' small wares ': pick out some examples from the list above.

Derby was selected by the old London Midland and Scottish Railway as their chief centre for locomotive building and repairing. One reason for this choice was the proximity of coal and iron ((*18*) *Where?*): another reason is apparent from the map on the left—(*19*) what is it?

British Rail have their technical headquarters and main engineering workshops in Derby, and the town is Britain's foremost engineering centre. The world-famous Rolls Royce works alone employs 20 000.

Derby's position as a route centre is emphasized in this map. (20) Using your atlas, decide (a) which river-valleys you would follow, and (b) the approximate compass directions taken, on leaving Derby by train for London, Manchester, Birmingham and Hull. (21) Next prepare a map similar to that on the right but adding information about these routes by inserting labelled arrows.

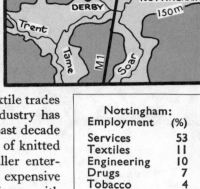

Position of Derby

The River Trent and its flood plain are narrow at Nottingham, where a ford existed in early times. The town grew up on a defensive site on a sandstone hill rising above the plain.

These figures show that the modern city has many more industries than the textile trades mentioned above. In fact the textile industry has suffered a loss of 10 000 jobs during the past decade as a result of low-cost imports, especially of knitted garments. The surviving firms are smaller enterprises which concentrate on making expensive clothing. Three famous *non*-textile firms with factories in Nottingham are Players (cigarettes), Boots (drugs and cosmetics) and Raleigh (cycles).

Nottingham: Employment	(%)
Services	53
Textiles	11
Engineering	10
Drugs	7
Tobacco	4
Coalmining	2
Others	13

Raleigh is the world's largest manufacturer of cycles and cycle components. The firm also makes wheeled outdoor toys, prams, pushchairs and gaming machines. Nottingham is also the main administrative, commercial and shopping centre of the North-East Midlands: (22) how is this fact reflected in the employment table?

Since 1960 coalmining in this region has suffered great changes of fortune. Derbyshire was especially hard-hit, for many older collieries there became exhausted and were closed. Some mining districts, e.g. the Erewash Valley, lost all their pits and the need for alternative employment became urgent. Industrial estates were built adjacent to the M1 at Worksop and Alfreton, and many Derbyshire miners now travel to collieries in Notts. Here, too, some inefficient pits have been closed and others modernised. Altogether the region has lost 37 000 coalmining jobs.

Now all of the collieries are fully mechanized. Powerful cutting and loading machines and computer control systems make them highly productive. New seams are being tapped and the coal output of 29m tonnes p.a. forms a rich national asset. As well as the deep shaft mines there are some large opencast workings. These, by levelling old spoil heaps, are transforming the landscape: one such project created the Shipley Lake park near Ilkeston.

56 kilometres from the sea . . .

CHAPTER 9

The Lancashire and Greater Manchester Industrial Region

THE 6155-tonne vessel S.S. *Geologist* is discharging a cargo of raw cotton at Manchester docks. Her voyage inland lay along the Manchester Ship Canal—opened in 1894 and costing up to date £40 million for construction and development. Why was this expensive link with the sea constructed ? Why was a port for large ocean-going cargo boats built here at the western foothills of the Pennines ? The answer is largely contained in the words *coal* and *cotton*.

On page 109 we saw that the Lancashire coalfield occupies a position on the western flanks of the Pennines similar to that of the Northumberland and Durham, and the Yorks., Notts. and Derby fields to the east. One important difference, however, is that in Lancashire the concealed coalfield lies too deep to be workable. (*1*) The reason for this is shown in the cross-section diagram opposite: write it down in words.

In the close-packed industrial towns on this coalfield between the Ribble and the Mersey (map overleaf) are the spinning, weaving and finishing mills of Britain's largest cotton industry. Cotton first came to Lancashire some 300 years ago. For centuries previous to its introduction, woollen and linen cloth had been manufactured there. The arts of spinning and weaving

128

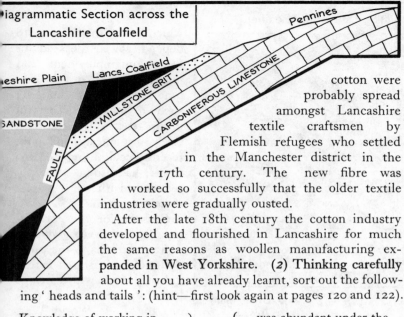

Diagrammatic Section across the Lancashire Coalfield

cotton were probably spread amongst Lancashire textile craftsmen by Flemish refugees who settled in the Manchester district in the 17th century. The new fibre was worked so successfully that the older textile industries were gradually ousted.

After the late 18th century the cotton industry developed and flourished in Lancashire for much the same reasons as woollen manufacturing expanded in West Yorkshire. **(2) Thinking carefully** about all you have already learnt, sort out the following ' heads and tails ': (hint—first look again at pages 120 and 122).

Knowledge of working in textiles		was abundant under the textile towns
Experiments with the new fibre		was available for bleaching and dyeing
Numerous swift-flowing streams	**?**	were unhindered by guild restrictions
Coal to drive steam-engines		was already traditional in Lancashire
Soft water in moorland streams		existed for water-driven machinery

Important inventions of spinning and weaving machinery in Lancashire stimulated the country's growing cotton industry. The damp Lancashire climate also aided the industry's original growth, for cotton fibres break in dry air. Today, however, the humidity in the mills is controlled by machinery.

By the late 19th century coal and cotton dominated the region. (*From the map and diagram overleaf (3) make a list of* (a) *the main spinning towns and* (b) *the main weaving towns.* (4) *Explain why these towns grew up on the flanks of the Rossendale moors and the Pennines.*) Yet the very prosperity of these two industries, on

SOME LANCASHIRE TEXTILE MACHINERY INVENTIONS

1733 Kay's flying shuttle	1779 Crompton's mule
1768 Arkwright's spinning frame	1785 Cartwright's power loom
1770 Hargreaves' spinning 'jenny'	

The Lancashire Coalfield

▲ Old Weaving Towns
● Old Spinning Towns
○ Old Textile Finishing Towns
▽ Other Main Towns

which whole towns came to depend, sowed the seeds of future unemployment and misery. Since 1913 the cotton industry has suffered such a long and painful decline that today it is relatively insignificant, and more recently coalmining in Lancashire has also been sharply reduced, with only 9 pits remaining open.

Reasons for the contraction in coal were given on p. 82. Threats to the prosperity of Lancashire's cotton industry began with a serious loss of overseas markets during the First World War. Cut off from their normal sources of supplies, many countries, including India, Japan, China and Brazil, set up factories of their own and were soon supplying cheap fabrics to many of Britain's former buyers.

India, especially, had the advantage of very cheap labour, local supplies of raw cotton, and a huge home market for cotton cloth. Furthermore, by starting so much later, Indian factories were able to install the latest automatic machinery and to use efficient methods of production. Many Lancashire factories, by contrast, were already antiquated in the 1930s.

To meet this challenge the Lancashire cotton industry has been severely pruned and modernised: between 1960 and 1975 the number of spindles was cut from 17.5m to 2.25m, and looms from 270 000 to 50 000. Much new automatic equipment has been installed and—a comparative novelty in Lancashire—it is used on a two- or three-shift system. The modern, streamlined textile industry which survives is very different from that of former times. Old-type cotton woven fabrics have largely given way to new blends of both natural and man-made fibres, and the mills now concentrate on producing high-quality clothing.

This transformation led to marked changes in the former cotton towns. "The mill chimneys, which stand like clusters of grave-

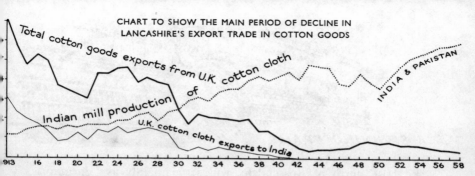

Total cotton goods exports from U.K. cotton cloth

Indian mill production of

U.K. cotton cloth exports to India

INDIA & PAKISTAN

1913 16 18 20 22 24 26 28 30 32 34 36 38 40 42 44 46 48 50 52 54 56 58

stones, gradually lost their smoke in one of the most rapid run-downs of a single industry the country has seen. They still make cotton in these towns, but it is a mere trickle compared with earlier times." (*The Financial Times.*) Typical in this respect is Bolton (152 000), once the hub of the world's fine cotton spinning. Now many of the once famous mills are filled with new light industries (*see right*), and already by 1967 the number of Bolton residents employed in textiles was surpassed by those in engineering. Bolton has also

> ### TYPICAL PRODUCTS IN FORMER COTTON MILLS
>
> Electrical goods, rainwear, plastics, automobile parts, cigarettes, frozen poultry, domestic appliances, cosmetics, Christmas puddings. These new industries provide Lancashire with about 155 000 jobs. The cotton industry, by contrast, has cut its labour force since 1945 from **320 000 to 80 000.**

become Britain's main centre of the mail-order industry: former mills make useful warehouses and the town is well placed as a distribution centre on the North's new motorway system. (*Map, p. 134.*)

In an effort to attract new industries the old cotton towns are providing new roads, clearing away slag-heaps and slums and re-building their town centres, yet unemployment remains well above the national average and young people continue to leave. It seems likely that in future many of these towns, especially

Below is a section through the Lancashire industrial area along line *XY* on the map opposite. ake a *large* copy of it and add a suitable title. Name the three cities shown and add the following bels in their correct places:

SPINNING	LIMESTONE	ROSSENDALE FOREST	FAULT
WEAVING	MILLSTONE GRIT	RIBBLE VALLEY	SANDSTONE
MARKETING	COAL MEASURES	MERSEY PLAINS	FINISHING

A SPINNING FRAME

those north of Rossendale, will serve mainly as 'dormitories' for people travelling daily to work in the more prosperous Manchester conurbation.

Another important industry in the North-West is the manufacture of textile-making machinery, i.e. TEXTILE ENGINEERING. Ironically, it was the export of British machines to countries such as China, Japan and India that helped bring about the serious depression in the cotton industry in the 1920s.

Large numbers of such machines, but especially those for cotton cloth making, are produced in Greater Manchester (14), Colne (6), Rochdale (5), Oldham (4), Blackburn (4), Bury (3), Bolton (2), Preston (1), Accrington (1), Burnley (1) and Stockport (1). The figures in brackets show the number of firms in each town which make machine tools as well as a great variety of textile machines. (5) Use them to draw a map showing the relative importance of these towns as textile-engineering centres. Show each firm by one dot, e.g. ● PRESTON. Mark in the coalfield and give your finished map an appropriate heading.

The large-scale growth of textile engineering in Lancashire was aided by (a) big demands for machines from the world's expanding textile industries ((6) *Can you suggest a likely importer of British machines other than those countries mentioned above?*); (b) abundant supplies of local coal ((7) *needed for what?*); and (c) nearby iron and steel production in the Midlands (see page 146) and Sheffield (page 124). Despite the set-backs of the depression years Lancashire textile engineering is once again prosperous. (8) Can you suggest why the industry has revived? (*Hint: synthetics.*)

Developing from the earlier textile-machine making industry, other types of engineering grew up in Lancashire. Now, in the old 'cotton belt', more workers are employed in engineering than

STEAM TURBINE ROTOR

in the more famous textile mills. Especially important are:—

HEAVY ENGINEERING	ELECTRICAL ENGINEERING
i.e. producing goods such as . . .	i.e. producing goods such as . . .
Structural steelwork	Transformers, switchgear
Locomotives (steam and diesel)	Electron microscopes, batteries
Stationary steam engines	Welding equipment and electrodes
Machine tools, e.g. hammers	Railway signalling equipment
Vehicles, e.g. Leyland trucks	Radar and X-ray appliances
Mining and colliery equipment	**Calculating machines, computers**
Hydraulic machinery	Electric motors and locomotives
Machinery for flour milling	Meters and laboratory instruments
Gas and oil engines	Turbo-electric propulsion for ships
Steel rolling-mill equipment	G.P.O. automatic telephone stations
Wire ropes, cables and chains	Industrial heating equipment

(9) Which of the goods in the above lists do you think were not made until the present century?

Heavy engineering has suffered badly during the past decade as a result of successive trade recessions. The heavy electrical industry has also been rationalised ((10) *Explain*), with the loss in the early 1970s of 10 000 jobs. One result is that many premises formerly used as engineering workshops in the Trafford Park estate (*see photo*) are now occupied by distributive, freight and storage firms. Unfortunately warehouses and depots are not **labour intensive,** i.e. they do not provide many jobs. This problem has been eased by the growth of service industries and light engineering. Ferranti, for example, now have their H.Q. near Oldham and six factories in Manchester. Products include micro-circuits and other components for military equipment.

Part of the Trafford Park industrial complex, Manchester.

Manchester (506 000) is the nerve-centre of the Lancashire Industrial Region and so has been profoundly affected by the economic changes referred to in the last few pages. The city first gained fame and prosperity as the commercial focus of the textile industry. The map on page 130 shows that Manchester is ringed to the north and east by textile towns. Most of the marketing and financial transactions for the textile industry are organised in Manchester's business houses and banks.

Now the city is more concerned with engineering and electronics and other new growth industries, no fewer than 157 000 non-textile jobs having been created there since 1945. In the port area adjoining the Manchester Ship Canal lies the Trafford Park Industrial Estate, where there are some 200 factories, warehouses and depots. Trafford workers are engaged in such varied industries as engineering, detergents, petro-chemicals, paper and processed foods. Greater Manchester has a population of 2.7 million, and a total of 19 million—one-half the population of England and Wales—live within a radius of 160 kilometres. This fact, coupled with the position of Manchester as a focal point of motorway routes (*see map*), has encouraged the growth of many service and distributive industries. Thus fleets of Manchester-based lorries deliver processed foods to supermarkets throughout the North-west, restocking from depots in Trafford Park and Middleton. Manchester docks also handle trade from as far afield as Tyneside, Teesside and West and South Yorkshire. (*11*) Which motorway facilitates this trans-Pennine commerce?

The international importance of Manchester is indicated by the

fact that the port operates a weekly trans-Atlantic container service to the St. Lawrence and Great Lakes. Furthermore Manchester's international Ringway Airport is the second in Britain by volume of goods handled. Manchester (with Liverpool) has also become Britain's largest centre for office administration outside London, and has a well-known and expanding university.

This motorway intersection of the M62, M61 and the A580 symbolizes the great changes in communication which are currently taking place in Lancashire. "Access to motorways has virtually dictated the pattern of recent manufacturing and distributive growth and few projects of importance are now far removed from a motorway link." (Financial Times.) Notice that the M6 runs like a spine through the North-West. (*12*) Which of the region's four New Towns does not lie quite close to this traffic artery? (*13*) Which motorway links this town to the M6?

Skelmersdale is a 'boom town'—the fastest growing New Town in Britain. Its 1964 population of 12 000 had doubled by 1972 and is expected to reach 60 000 by 1990. Industries there include engineering and synthetic rubber production. The brand new town of *Central Lancashire* is being built between Leyland, Chorley and Preston. This is the biggest project of its kind yet attempted in Britain and the population of the area is planned to rise from its present 250 000 to more than 400 000 during the 1990s.

With the M6 now on its doorstep the old coalmining town of Wigan has also been able to transform itself. Wigan once had 600 coalpits and was notorious for its industrial grime and squalor. Now only two mines remain open, the spoil heaps have vanished —the former 'Three Sisters' tips covering 120 hectares were bull-dozed to make a park—and former miners are employed in a wide range of new industries, including food-processing, mail-order trading, retail distribution and engineering. (*14*) Suggest how each of these is favoured by proximity to the M6.

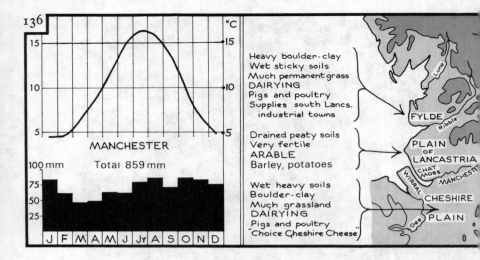

CHAPTER 10

The Lowlands flanking the Pennines

SOME ten million people are packed into the industrial districts of the Pennine coalfields. The dense clusters of population living in West and South Yorkshire, Greater Manchester and the Derby–Nottingham area rely on foodstuffs brought from elsewhere. As the industrial cities grew they needed to draw increasingly upon the agricultural supplies of the surrounding countryside. Fortunately, the lowland districts on both flanks of the Pennines include some of Britain's best farming land. The type of farming depends on several factors:—

(a) *Soils* are fertile on the New Red Sandstone which underlies most of these lowlands. Large areas, too, are covered by a mantle of glacial drift: some of the sandy deposits are sterile, but there is much extremely fertile boulder clay. (If you are not sure how these glacial deposits were formed, revise pages 35–36.) Land use varies with the type of soil, as for example:

(*i*) on the reclaimed peat of Chat Moss large quantities of vegetables are grown;

(*ii*) the reclaimed alluvial land between the lower Ouse and Derwent is very favourable for potatoes and sugar beet;

(*iii*) potatoes, peas and carrots thrive on the light, sandy, glacial soils in the northern part of the Vale of York;

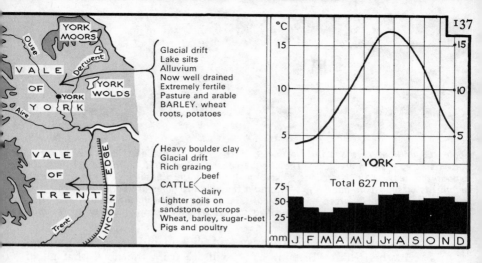

Map labels:
YORK MOORS · Ouse · Derwent · VALE OF YORK · Aire · YORK WOLDS · YORK · VALE OF TRENT · Trent · LINCOLN EDGE

Glacial drift
Lake silts
Alluvium
Now well drained
Extremely fertile
Pasture and arable
BARLEY. wheat
roots, potatoes

Heavy boulder clay
Glacial drift
Rich grazing
CATTLE⟨beef / dairy⟩
Lighter soils on
sandstone outcrops
Wheat, barley, sugar-beet
Pigs and poultry

°C — YORK — Total 627 mm

(*iv*) dairying (now mainly for milk rather than cheese) is the main activity wherever heavy boulder clay produces wet, sticky soil unsuited to ploughing.

(*b*) *Climate* affects the use made, even of similar soils, in different districts. E.g. a clay usable only for pasture in rainy Lancashire could be ploughed in the drier Vale of York.

(*c*) *Location.* E.g. around great industrial cities like Liverpool, Manchester and Leeds are many market gardens, the produce being sent daily to the near-by city markets.

(*1*) Study the information given on these two pages and then write notes comparing and contrasting farming in Cheshire and Nottinghamshire. Suggest reasons for the statements you make. Include in your account all the following words:

moist	glacial drift	oats	dairying
cereals	beef cattle	sunnier	market gardening
drier	potatoes	equable	sugar-beet

Farm Facts	Cheshire	Nottinghamshire
Total area (in ha) under		
crops and grass	190 165	156 919
Permanent grass	91 119	38 464
Rotation grass	62 565	29 428
Crops	36 415	88 996
including		
wheat and barley	33 201	74 689
sugar-beet	55	7542
Dairy cattle	147 459	32 535
Other cattle	112 914	72 955
Pigs	106 101	98 775
Sheep	23 216	99 507

Some 200 million years ago the Northwich district of Cheshire must have looked very like th[is] salt-encrusted plain in Nevada. The climate became so arid that an inland salt lake complet[ely] evaporated, leaving thick deposits of salt on its bed. The Cheshire salt now lies beneath la[ter] sediments of sandstone and clay. To obtain the salt two holes are bored through the overly[ing] rocks. Fresh water is forced down one hole and brine pumped up the other. The brine is th[en] evaporated and the salt residue is purified for use as a basic raw material in the mid-Mers[ey] chemical industry.

As well as chemicals, soap and glass manufacture is important in this part of Britain. You will see from the notes that all of these industries are closely connected.

MIDDLE MERSEY REGION: MAIN INDUSTRIES

Industry and Main Towns	Raw Materials Needed	Source
CHEMICALS Northwich Widnes Runcorn Winsford	Common salt (sodium chloride), for making chlorine, caustic soda and hydrochloric acid	Cheshire salt beds
	Coal and Electricity	Lancs. & other Pennines coalfields
SOAP Port Sunlight Warrington Widnes Manchester Liverpool	Fat (tallow from sheep and cattle)	Australia Argentina
	Vegetable oils e.g. palm kernel oil ground-nut oil copra oil	Mostly tropical Africa Processed in Liverpool
	Caustic soda Common salt	Cheshire salt beds
	Coal	Lancs. coalfield
GLASS St. Helens	Pure sand (silica)	Shirdley Hill (Lancs.) and import[s]
	Sodium carbonate	Cheshire
	Limestone	Pennines
	Coal	Lancs. & other Pennines coalfields
OIL REFINING AND PETROCHEMICALS Stanlow Ellesmere Port Eastham	Crude petroleum Refined petroleum products	Imports, mostly from Middle East and Venezuela Local refineries, and by pipeline from Milford Haven (p. 97)
VEHICLES (see also notes, right) Speke ⎫ Halewood ⎬ Liverpool Ellesmere Port ⎭	Steel	English Midlands and imports
	Glass	St. Helens
	Fibres	Lancashire textile towns
	Rubber	Leyland

The Site of Liverpool

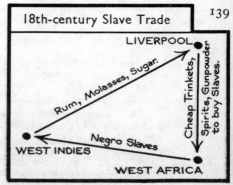

Labels: D', Peat Bog, Ridge, The Pool, D, Peat Bog, Mersey, SANDSTONE MOUND, RIVER MARSH, OLD ROAD

18th-century Slave Trade

LIVERPOOL

Rum, Molasses, Sugar.

Cheap Trinkets, Spirits, Gunpowder to buy Slaves.

Negro Slaves

WEST INDIES

WEST AFRICA

Much of south Lancashire was once a marsh. (2) From the map above, explain why Liverpool grew up at that particular site on the Mersey estuary. The original small port, trading mostly with Ireland, grew up on a tidal inlet called the Pool. Liverpool first began to prosper in the 18th century, when the city's merchants accumulated vast fortunes in the notorious, but highly profitable, Slave Trade Triangle. ((3) Why this name?) In spite of the abolition of the slave trade in 1807, Liverpool's commercial links with America remained: she rapidly became a major importing centre for raw cotton and foodstuffs from the New World.

Liverpool's modern port has 200 hectares of docks lying along 50 km of the Mersey's north bank. (*From D to D' in the sketch-map:* (4) *what natural conditions favoured dock excavation here?*) There are now also 15 km of docks on the Birkenhead side of the estuary and the two cities are joined by rail and road tunnels running beneath the Mersey. (5a) Draw a labelled sketch-map to show the following developments in Liverpool–Birkenhead:—
(i) A new container port at Seaforth handles bulk carriers of up to $\frac{1}{4}$ m. tonnes and is served by an inland container base at Orrel; (ii) Bootle has a new freightliner terminal; (iii) Birkenhead's shipyards are being modernised; (iv) a new urban 'Loop-Link' railway is being built for commuters into central Liverpool.

Vehicle manufacture is Merseyside's latest growth industry. Vauxhall at Ellesmere Port, and Ford and Standard-Triumph at Liverpool between them employ 30 000 workers. These firms, guided by Government policy to relieve unemployment, came to Merseyside during the 1960s. Now the car and vehicle component factories form one of Europe's largest vehicle manufacturing centres. Older and smaller works at nearby Crewe, Sandbach and Leyland make heavy commercial vehicles, e.g. trucks, dumpers and fire-engines.

Middle Mersey and Cheshire Salt District

Main Imports			LIVERPOOL—BIRKENHEAD	Main Industries
Cotton	Flour	Vegetable oils	Flour milling	Manufacture of:
Meat	Hides	Sugar	Sugar refining	Soap and Margarine
Grain	Tobacco	Metal ores	Tin refining	Chemicals
Wool	Timber	Rubber	Ship building	Cattle-cake
Eggs	Tea	Fruit	Ship repairing	Cigarettes

(5b) Trace some of the links between the chief imports and industries of Liverpool—Birkenhead in this table.

Liverpool is one of Britain's main ports for the import and export of goods other than fuels: in particular it handles one-seventh of the country's total exports. At first Liverpool served as a port merely for the lowland of South Lancashire. Today its **hinterland** (i.e. the area behind a port with which it has close commercial dealings) is very extensive. The main industrial areas which Liverpool serves as a port are shown on this map, and so are the goods handled from Stoke and the Midlands.

The Hinterland of Liverpool

(6) Suggest, giving reasons, raw materials and products passing through Liverpool to and from the other cities shown.

The extension of Liverpool's hinterland into Yorkshire was made possible by early canal and rail links across the Pennines, and is now maintained by modern roads like the M62. (*See p. 60.*) Whether goods for export cross the Pennines depends largely on their final destination, for the industrial cities on the eastern flanks of the Pennines have a second and more direct access to the sea at HULL, on the Humber estuary. Indeed, it was the development of wool and iron manufacturing in West Yorkshire which stimulated Hull's growth. Important, too, was the decay of ports like Selby farther up the estuary as the size and draught of ships steadily increased.

Look now at the site of Hull on the map on the next page. (7) In what way does it resemble that of Liverpool? (8) What do you notice, too, about the depth of water in the Humber at Hull? (9) How do you explain this depth? (*Hint: read page 24 again.*) Rivers such as the Aire, Calder and Don, which converge on the Humber, have been canalised to facilitate barge traffic.*

Much cargo handled at Hull is distributed to the great industrial cities of West and South Yorkshire. Cargoes are also assembled for despatch overseas, especially to Scandinavia and the Common Market. Thus Hull is a **bulk-breaking point**. (10) *Explain this term.* The three ro–ro ferries which operate from the port (to the

* Most of the 4.5m tonnes of cargo carried each year consists of coal and coke on its way to power stations, or to Goole for export. (*Map, p. 119.*)

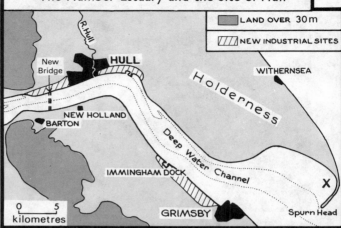

The Humber Estuary and the Site of Hull

LAND OVER 30m
NEW INDUSTRIAL SITES

The Humber is the biggest but least developed estuary in Britain. Future development hinges mainly on improved access by road: thus a suspension bridge has been built (*see map*) and the M62 extended right into Hull (*map, p. 281*). (*11*) Why is Humberside particularly well situated to gain from Britain's entry into the Common Market? (*12*) With which other British estuaries does it compete in this respect? There ample room for industrial expansion on Humberside, especially downstream from Hull (*see map*). Plans exist for a bulk-carrier dock at point X on the map: what natural advantages would favour this site?

Netherlands, Belgium and Sweden) carry 500 000 passengers each year as well as merchandise.

Hull is Britain's largest fishing port, Hull and Grimsby together landing over half of Britain's supplies of fish. The fishing industry, however, faces problems now that British fishing vessels are barred from Icelandic waters. The port is therefore increasingly reliant on new growth industries, notably caravans ($\frac{1}{3}$ of U.K. output), drugs and electrical engineering.

In recent years an oil refinery has been built and a whole new range of chemical industries developed between Grimsby and Immingham Dock (see map). The factories here can easily import raw materials, and discharge effluent into the tidal channel. The massive supplies of pure fresh water needed in the chemical works are obtained locally from five large wells. (*13*) From which rock stratum is this water extracted? (*Hint: pp. 30–31 and map, p. 29.*) Grimsby has processing and quick-freezing works for fruit, vegetables and fish.

MAIN INDUSTRIES IN HULL

Flour milling
Saw milling
Food processing
Oil seed crushing
Animal feedstuffs manufacture
Engineering (e.g. making machines to crush oil seeds and process the oil)
Petro-chemicals.

HULL—MAIN EXPORTS

Chemicals and fertilizers from Humberside
Petroleum (coastal shipments) and bunker fuel from Humberside
Machinery and vehicles from Midlands
Iron and steel goods from South Yorks.
Foodstuffs from local factories
Textiles from Gtr M'ch'st'r and W. Yorks.
(*14*) Use this information to make a map showing the hinterland of Hull, similar to that for Liverpool shown opposite.

MAIN IMPORTS

Petroleum	Fruit and vegetables
Timber	Molasses and sugar
Grain and flour	Oilseeds and nuts
Iron and steel goods	Textile fibres
Fish landings	Other foodstuffs

These chalk cliffs at Flamborough Head are more than 300 kilometres from the better-known white cliffs of Dover. Yet they remind us that the scarplands of chalk, clay and limestone which make up south-east Britain extend as far north as Yorkshire. Here, between the Vale of York and the North Sea, they form the North York Moors, the Vale of Pickering and the Yorkshire Wolds.

The North York Moors

A Jurassic Sandstone escarpment. The undulating moorlands are covered with coarse grasses and heather on which sheep and cattle are reared. Many dip-slope rivers have cut deep, fertile valleys: e.g. Eskdale. Here, on alluvial soils, root crops and cereals are grown.

The Cleveland Hills to the north once produced much iron ore ((*15*) *For what nearby industrial centre?*) but the mines closed in 1964.

The Yorkshire Wolds

The dip-slope of the chalk escarpment of the Wolds has been covered in many places by patches of fertile boulder clay. The presence of this clay, aided by scientific farming, has allowed widespread arable cultivation in spite of the thin soils. The ground is manured by the folding of sheep. Wheat, malting barley and sugar-beet, together with turnips and swedes (for the large numbers of sheep) are the chief crops.

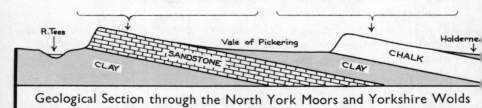

Geological Section through the North York Moors and Yorkshire Wolds

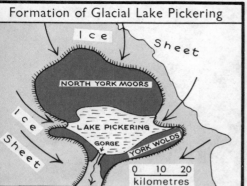

Formation of Glacial Lake Pickering

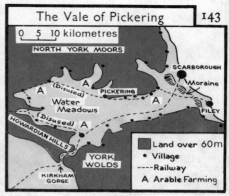

The Vale of Pickering

The ice-sheet (page 32) which dumped boulder clay on the eastern parts of the Yorkshire Wolds also sealed the gap between the North York Moors and Flamborough Head. At the same time the western end of the Vale of Pickering was blocked by ice occupying the Vale of York.

The River Derwent was thus prevented from reaching the North Sea along its old course at Filey Bay, and glacial 'Lake Pickering' was formed.

The water eventually spilled over the western edge of the lake and cut a deep **glacial overflow channel** at Kirkham Gorge. (*16*) Why did the River Derwent continue to drain westwards to the Ouse and the Humber after the final melting of the ice-sheets?

The silts which were laid down on the bed of the former lake now form fertile soils, but the central part of the Vale is low-lying and liable to flooding. (*17*) How has this affected (*a*) the position of roads, railways and villages, and (*b*) the agricultural activities in the Vale of Pickering?

Where the boulder clay and gravel patches of the Holderness Plain have been drained, wheat and roots are grown and cattle reared.

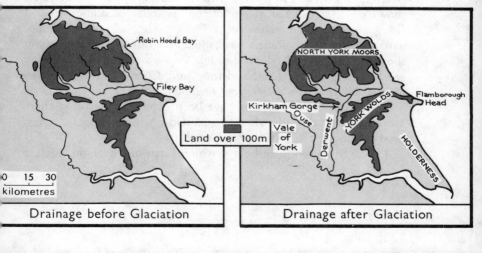

Drainage before Glaciation

Drainage after Glaciation

CHAPTER 11
The English Midlands

THE TRIANGULAR part of Britain between the Welsh Mountains, the Southern Pennines and the Jurassic Escarpment is usually called the Midland *Plain*, but the general flatness of the landscape is broken in places by hilly districts. The photograph shows the Malvern Hills—one of the 'islands' of higher ground rising abruptly from the plain. These 'islands' consist of old, hard rocks such as shale, quartzite and granite. The diagram shows how they protrude through coal-bearing strata and the sediments of New Red Sandstone and Clay. ((*1*) *What other ancient hills can you detect from the maps?*)

Where the Coal Measures have been exposed by erosion there are several small but valuable coalfields, which altogether now produce about 7% of the total British coal output.

The reddish sands and clays which form much of the surface

144

Midlands: Physical and Main Towns

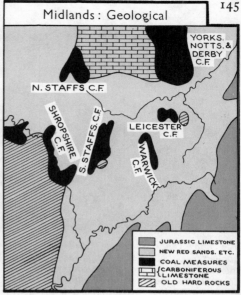

Midlands: Geological

layers of rock in the Midlands were formed long ago under desert conditions similar to those found in the Sahara. The clays were once silts on the beds of salt lakes like that shown on page 138; the sandstone was built of wind-blown dunes like those accumulating today south of the Atlas Mountains (see pages 10 and 20). Amongst the deposits which crystallised out on the beds of the ancient salt lakes was calcium sulphate (gypsum). This salt is today quarried from the New Red Sandstone near Tutbury, in Staffordshire. It is used to make plaster of Paris and plasterboard—an important modern building material.

Through millions of years the climate and scene have gradually changed so that today " here is the red English plain . . . extending along the valleys of the Severn and the Stratford Avon, the England of Shakespeare's country, with its roads fringed with great trees, with its red ploughed fields making the green of the hedges seem a deeper tint, with innumerable villages of red brick or black and white thatched cottages set along twisting roads ".[1]

[1] Trueman, *Geology and Scenery in England and Wales* (Pelican).

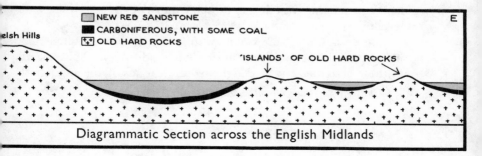

Diagrammatic Section across the English Midlands

For many centuries the Midlands have been famous for their metal industries. The first successful use of coke for smelting iron ore was introduced on the Shropshire coalfield in 1709 by Abraham Darby. Other important inventions followed, and the photograph shows the first iron bridge, built across the River Severn in 1779. (Where exactly? Look at the map at top page 147.) Although the bridge is no lon_ used by traffic it still stands as a monument to _ engineering skill of the pioneer Midlands me_ workers. A mile away at Coalbrookdale (_ home of Abraham Darby) the first iron ba_ and the first iron tramway lines were made.

At the time of these inventions iron smelters chiefly relied on ' black-band ' iron-ore deposits in the Coal Measures. The Midland fields were rich in such ores and iron smelting there spread **rapidly, especially in the area between Birmingham and Wolver-hampton.** Here the countryside was made so ugly by pit-heaps and drab, smoke-begrimed factories and houses, that the district was called the Black Country—the name by which it is still known.

Today very little iron smelting is done in the Black Country, for the coalfield is worked out, and the ironstone went long ago. "Thousands of acres of derelict land, unlike anything else in England, a black, tormented Sahara, mark where the shallow coal-pits were, and deep, water-filled holes where the brick-clay pits were opened up. This is the aftermath of centuries of robber-exploitation of the natural resources of the earth, regardless of all the consequences. Today . . . heavy excavating and soil-shifting machinery . . . is being used to reclaim this derelict land, by moving vast heaps of spoil into equally vast pits, and levelling the landscape out."*

The decline of Black Country iron *making* was accompanied by the growth of a multitude of iron *using* industries. At the heart of these activities lay Birmingham—the capital of the industrial

* *Chilterns to Black Country*, ed. Grigson (Collins).

Midlands. (2) From the figures draw a chart to show the increase in population in Birmingham between 1650 and 1971.

Metal trades have a long history in Birmingham. The 16th-century traveller Leland found " that a great part of the towne is maintained by smiths who have their iron and sea-cole out of Staffordshire ". A generation later Camden described the city as " full of inhabitants and resounding with hammers and anvils, for most of them are Smiths ". Then in 1761 two men of great genius—a businessman, Matthew Boulton, and James Watt the inventor —opened a factory in the Soho district of Birmingham to produce high-quality metal wares. Their chief products were steam engines, which provided a great new source of power to drive machines. Engines from the Soho works did much to mechanise and transform British industry, and the Black Country became Britain's chief centre of heavy industry— a position kept for over a century.

Throughout the 19th century orders from all parts of the world poured into Birmingham for steam engines and machinery of all kinds, and metal working and engineering still predominate among the city's many industries. Typical Birmingham products include brassware, electroplated and galvanised steel, motor cycles and vehicles.

Birmingham lies on one of the 'islands' of old hard rocks mentioned on p. 145. Lack of river transport in this hilly area originally hampered the city's growth, but expansion became phenomenal as transport facilities improved in modern times. (See below.)

POPULATION OF BIRMINGHAM	
1650 . .	1 500
1700 . .	15 000
1811 . .	85 755
1831 . .	142 251
1851 . .	232 841
1871 . .	343 787
1891 . .	478 113
1911 . .	840 202
1931 . .	1 002 603
1951 . .	1 112 685
1971 . .	1 013 366

The West Midlands Conurbation

Birmingham and the other towns shown in this map have grown to such an extent that they now merge together with no noticeable break. An almost continuous built up area of this extent and nature is called a **conurbation**. From the map (3) estimate the area in square kilometres of urban land within the Birmingham-Wolverhampton conurbation. This region contains 5.18 million people and Birmingham is Britain's second largest city. Further growth of the conurbation is being limited by planning authorities ((4) Why?) and overspill population from Birmingham is being re-settled in the New Towns indicated on the map. (5) Name them.

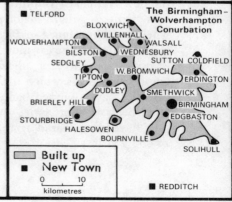

Copper tubes, locomotives, rubber tyres, cycles, chocolate, lead pipes, railw
wagons, car upholstery, plastics, aluminium castings, springs, watches, gold chair
rifles, machine tools, British Leyland cars, jewellery, bronze ware, kitchen utensi
commercial vans, glass, buttons, nickel plate, electrical plant, steel tubes, bedstead
nuts and bolts, silver plate, furniture, pins, galvanized iron, sewing machines, bra
ware, weighing machines, wireless apparatus, scientific instruments, carburetto
plastics, car engines, brass components, transmission shafts, copper radiator valve
vehicle bodies and sub-frames.

Some of the multitude of goods manufactured in the West Midlands are shown above. (6) Sort them into lists under the following headings: (1) METAL-WORKING TRADES, (2) GENERAL ENGINEERING, (3) AUTOMOBILES, and (4) MISCELLANEOUS. (7) Put an asterisk against those goods in your lists which are valuable but of small bulk, and require relatively little raw material for their manufacture (e.g. *watches**). The West Midlands towns specialise in making goods of this kind: but their heavy industries declined as the 'black-band' iron-ore ran out in the late 19th century. Remoteness from the sea made the import of ores and the export of bulky products a costly business. Prices of goods like watches, however, are little affected by transport costs, and in this type of commodity the West Midlands could compete on equal terms with other industrial districts. Such products are still typical of the region, even though the disadvantages of distance have largely been resolved by modern road and rail facilities.

Before modern roads and railways were built, the Midlands were criss-crossed by an intricate network of canals which focussed on Birmingham. By providing cheap transport for imported ores the canals favoured the growing metal trades of the West Midlands. Today the canals carry only a tiny fraction of Midlands freight, most of which travels by road. (*See photo.*) (8) *For what goods do canals still offer certain advantages?*

The motorways which slice through Birmingham give the city a dynamic appearance. So does the busy international airport and the new National Exhibition Centre at Bickenhill (*see map*). Yet these tables highlight major problems. We have noted the overwhelming importance in Birmingham of engineering and vehicle manufacture. The car industry alone gives direct employment to 87 000—i.e. 1 in 7 of all workers. A further 260 000 work in engineering firms which make vehicle components. Since 1970 car sales have slumped badly and thousands of car workers have often been on short time. Meanwhile traditional manufacturing jobs have disappeared from central Birmingham at the rate of 10 000 a year.

Motorway links
at Birmingham

Airport

B'HAM

M6

M6

M5

A45

Proposed
Motorway

National Exhibition
Centre

0   20
     km

The M1 near Coventry. Increasing numbers of lorries pick up cargoes at and container depots (inland ' ports '), where freight is assembled for despatch. Such depots have proved to be the most useful new facility available to Midlands exporters since the invention of steamships. The map shows that Birmingham is especially well located in relation to the country's motorway network. (9)

Refer to p. 281 and state which motorway(s) a lorry with goods for export would follow on a journey from Birmingham to each of the following ports: (a) Liverpool; (b) London; (c) Bristol; (d) Hull and (e) Cardiff. (10) Of all possible routes, why do you think that between Birmingham and London was chosen for Britain's first motorway? (*Hint: see pages 256–7.*)

" The city was one of the first to raise the alarm about the disturbing and accelerating decline of inner urban areas . . . In the boom years of the motor industry the city suffered from labour shortages and there was a logic to the Government's policy of encouraging firms to move to the New Towns . . . Ironically, the drift away from the central area was accelerated by the City Council's ambitious slum clearance and road building programme. Redevelopment took people to better housing in the suburbs but also removed the small back street premises that had acted as the seedbed for new firms and activities. The massive new highways that cut a swathe to the centre of Birmingham to make it so distinctively a car-orientated city left in their path pockets of dereliction and undeveloped sites. " (*The Financial Times.*)

Those left behind in the blighted inner ' core ' are mainly unskilled, poor and badly housed. One-third are coloured immigrants. £30 million is being spent in renovating the area and in building houses and small factory premises. It is hoped to attract back skilled workers and various light and service industries. Elsewhere in the city, engineering firms are turning to export sales of car components, and some alternative jobs are available in new chemical, processed food and distributive trades.

BIRMINGHAM: POPULATION CHANGE 1971–76

Area	1971	1976	Change	Change %
...e Area	329 300	291 800	− 37 500	− 11.4
...t of Birmingham District	767 760	770 000	+ 2 240	+ 0.3
...mingham District Total	1 097 060	1 061 800	− 35 260	− 3.2

The opening in 1772 of the Trent-Mersey Canal stimulated a rapid growth on the North Staffordshire coalfield of what is today the world's largest pottery industry. The details of the rise and importance of this industry are shown in the notes below.

Until recently the 'Potteries' was the grimiest industrial region in Britain, blanketed by a perpetual pall of smoke and choking fumes rising from

The North Staffordshire Coalfield

hundreds of bottle-shaped coal-fired kilns. The kilns, together with thousands of little brick houses of the pottery workers, were black with their own dirt of generations. Today the smog has gone, for modern kilns are heated by gas or electricity, and some works are re-housed on pleasantly laid-out sites like that shown on the right. Local coal, once the mainstay of 'bee-hive' kilns, now goes to electricity generating stations and blast furnaces. Furthermore the famous canal has ceased to be of importance to the pottery industry, for transport is now by road and rail.

THE NORTH STAFFORDSHIRE POTTERY INDUSTRY

Origin and Growth	The Industry Today
Soils are poor on the south flanks of the Pennines ((11) Why?) and for centuries farmers here eked out a livelihood as part-time potters, using local clays. Brushwood was used at first for firing the kilns but was in time replaced by coal mined in the vicinity.	Much high-quality pottery is exported and the industry is an important 'dollar earner'. Smoke and grime has diminished, for gas and electric kilns have now replaced 'bee-hive' coal-fired ovens.
The industry owes much to the initiative and genius of Josiah Wedgwood, who set up his factories in the Potteries in the mid-18th century. He took an active interest in the construction of the Trent-Mersey Canal and built up the world-famous pottery firm which still bears his name.	Local clays are still used for articles like drain pipes, tiles, teapots, chimney-pots, and 'saggars' (containers in which fine pottery is baked). For quality wares like bone-china and porcelain the following raw materials are imported: China clay (kaolin) from Cornwall; Ball clay from Dorset; Flint from Northern France; Bone from Argentina.
Before this canal was built Cornish china clay for making the finer pottery was brought up the River Weaver by barge and then taken overland by pack-horse. The canal also provided cheaper and safer carriage for fragile pottery and helped bring foodstuffs to feed the growing population in the Potteries.	The chief pottery districts of Stoke-on-Trent are: Tunstall Burslem Hanley } white earthenware Stoke — high quality china and porcelain Longton Fenton } cheaper-quality goods

(12) Which motorway runs just west of Stoke-on-Trent? Indeed the survival of the pottery industry in this region is a good example of geographical inertia. *(13) Explain this, after re-reading page 80.*

Other important industries in Stoke-on-Trent include iron and steel production, engineering and the making of Michelin tyres. There is an urgent need for yet more alternatives to pottery, for the modern highly mechanised kilns have trimmed their labour requirements by 17 000 since 1968. *(14)* Draw bar diagrams to illustrate the following main sources of employment in Stoke-on-Trent: *pottery 44 000; engineering 25 000; coal-mining 7000; tyres 8000.*

The new smog-free appearance of the Potteries is matched by changes affecting other industrial centres, many of which have already undergone a 'spring-clean' to remove the accumulated dirt of two centuries of coal fires. At the same time great mounds of colliery spoil have been bulldozed and replaced by attractively laid out housing estates. Many older, semi-derelict districts have been re-built and redeveloped. In West Bromwich, for example, the new Sandwell Shopping Centre (*photo below*) has 3 stores, 60 shops, a pub, a multi-storey car park and a bus station all within the same group of buildings. It is hoped that the provision of amenities such as these will help attract new industries into the region and lessen its dependence on the motor trade.

Much more can obviously be done to smarten up the West Midlands conurbation, but to minimise congestion most new factories and houses will have to be located outside the existing cities. Planners suggest that growth should take place along a SW–NE axis through greater Birmingham: i.e. *(15)* towards which two major estuaries?

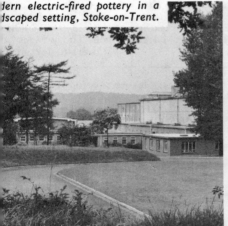

dern electric-fired pottery in a dscaped setting, Stoke-on-Trent.

New Sandwell shopping centre, West Bromwich.

The East Midlands

As yet there has been no large-scale growth of industry in the East Midlands comparable with that in the conurbations of Staffordshire. Compare this map with those on pages 147 and 150 and notice the absence of any large towns on the Leicestershire and Warwickshire coalfields. The coal-mining centres remain in most cases mere villages.

In the region around Northampton, however, a conurbation might appear in the future. A recent Government survey recommended a planned unification of Corby, Wellingborough, Northampton and Daventry. The East Midlands has important collieries, textile and leather industries and expanding engineering firms. (*See below.*) There are also large iron and steel works at Corby and Kettering. (*See page 158.*)

TOWNS AND INDUSTRIES IN THE EAST MIDLANDS

Engineering	Textiles	Leather: Boots and Shoe
Coventry:—Automobile and aircraft industries. The car industry developed from local cycle trades. Many skilled metal-workers and engineers from the nearby Black Country migrated here as these industries expanded. Rugby ⎱ electrical Loughborough ⎰ apparatus Leicester: heating and ventilating plant, textile machinery. Northampton: metal manufacture, especially of roller and other bearings.	Leicester, Loughborough, Coventry, Nuneaton. Specialist textile fabrics are made, e.g. hosiery, knitted suits, silk, ribbons, rayon. To the north lies the southern fringe of the Yorks., Notts. and Derby coalfield. (*(16) What other cities there produce similar textile specialities? See pages 126–7.*)	Leicester, Northampton, Wellingborough, Kettering, Rushden. Leather working grew here in the mark towns where plentif supplies of cattle-hid and sheep-skins we available. (*(17) Why? S map overleaf.*) Tannin extract came from o trees in the Midlan Plain. Today hides an skins are imported.

Three 'New' or 'Expanded' towns have recently grown considerably. *Northampton* (140 000), with new industrial estates, gained 10 000 extra jobs since 1968, mostly in engineering, warehousing and distributive trades (e.g. of beer and clothing). Original plans for expansion proved over-optimistic, but the town is expected to have 180 000 residents by 1990. *Milton Keynes* (74 000), a completely new settlement near Bletchley, has engineering, chemicals and printing industries, and is the home of the Open

University. Its population might reac 200 000 during this century. *Corby* (55 00 gained 12 000 residents after 1966, but no has difficulty in attracting more firm Miscellaneous light industries already the include potato crisps, trailers, lingerie ar tarmac, but the steelworks still provid 12 000 out of a total 27 000 jobs.

Present policy is to curb the rate a which 'New' and 'Expanding' towns ma grow. This is for fear of attracting tc many skilled workers away from existin cities. (*See also pages 149 and 185.*)

Examining nylon stockings for flaws before despatch from a Leicester factory.

A Northampton shoe factory worker attaching a heel.

Large reserves of untapped coal have been discovered in the East Midlands. One find is to the west of Coventry, where proved reserves amount to 200m tonnes. They include seams up to 7 metres thick of the prime Warwickshire Thick Coal. These important deposits can be worked from two existing collieries near Coventry. A second, and much larger find, lies in the Vale of Belvoir (*see map*). At least 500m tonnes of coal lies here in six seams at depths between 400 and 600 metres. This is the biggest single energy deposit so far discovered in the U.K.—bigger than the largest single North Sea oilfield. The National Coal Board wishes to produce 10m tonnes p.a. of this high-grade Belvoir coal. The scale of production will depend, however, on overcoming objections to mining from local residents and conservationists. (*18*) What are conservationists? (*19*) Why do they object? (*20*) Are their views (a) valid, (b) tenable, considering Britain's economic circumstances? (N.B. There are arguments both for and against.)

The 'ridge and furrow' pattern formed by the ox-teams of medieval ploughmen still shows in this Midlands pasture.

FARMING IN THE MIDLANDS

Away from the turmoil of the industrial cities one finds the other Midlands: the quiet streams, meadows and woodlands of peaceful farming country.

The rich red soils of the clay plain were first tilled by the Anglo-Saxons. They were **subsistence farmers,** growing crops and rearing livestock for their own needs. As they cleared the original forest the land was carved into great open fields. These were ploughed in such a fashion that the ground was thrown into a series of broad ridges and furrows.

Much of this land was heavy to plough and hard to keep drained, and so in later times the open fields were enclosed and converted into pasture for larger-scale, specialised sheep and cattle rearing. In some districts, with the use of powerful modern machinery, fresh changes resulted from the intensive 'ploughing up' campaign of the Second World War ((*21*) *Why was this necessary?*), but much of the Midlands is still under permanent pasture.

Both dairy and beef cattle are found in all districts, but the proportion of beef cattle increases towards the east, especially around Market Harborough and Melton Mowbray. Many specialist 'graziers' not only rear their own beef cattle but also buy in young hill cattle (e.g. from Wales) to fatten on the rich summer pasture.

Scale of diagram: 1 cm²: 5000 ha
(i.e. 1 mm²: 50 ha)
Each 'cow' symbol - represents 10 000 cattle

(22) **Construct similar diagrams for Hereford and Worcestershire and for Staffordshire, using the facts given below.**

(23) **Compare the three counties' farming and state (a) the broad similarities and (b) the detailed differences shown in the details given. Suggest reasons.**

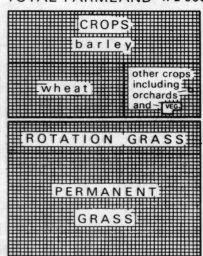

TOTAL FARMLAND 172 000 ha

CROPS
barley
wheat
other crops including orchards and VEG.

ROTATION GRASS

PERMANENT GRASS

DAIRY CATTLE

OTHER CATTLE

Around the West Midlands (Birmingham) conurbation there is a greater concentration on dairying and market gardening. (24) Can you explain this?

A highly specialised type of Midlands farming is found in the lower Severn Basin and particularly in the Vale of Evesham, which is famous for vegetables and fruit, particularly plums. Rich but rather lighter, loamy soils favour such farming here, and the markets of the industrial cities are close at hand. The climate— a slightly milder winter, earlier spring and longer summer than elsewhere in the Midlands—is also an advantage. Many orchards are away from the lowest parts of the Vale, the trees being planted on the lower hill slopes. (25) Can you explain this? (*Hint: pages 48-49.*)

	Hereford and Worcester	Staffordshire
Total area (in ha) under crops and grass	327 000	206 000
Permanent grass	151 000	120 000
Rotation grass	45 000	29 000
Crops	123 000	57 000
including:		
wheat and barley	74 000	44 000
orchards	12 000	400
vegetables	6 000	1 100
Dairy cattle	78 000	129 000
Other cattle	199 000	137 000

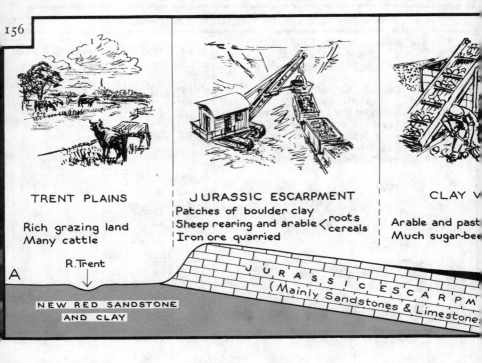

TRENT PLAINS

Rich grazing land
Many cattle

JURASSIC ESCARPMENT
Patches of boulder clay
Sheep rearing and arable < roots
cereals
Iron ore quarried

CLAY V

Arable and past
Much sugar-bee

R. Trent

A

NEW RED SANDSTONE
AND CLAY

J U R A S S I C E S C A R P M
(Mainly Sandstones & Limestone

CHAPTER 12

Lincolnshire and The Fens

THESE MAPS and diagrams show all the main facts about the geography of Lincolnshire. (1) Study them carefully and then write a geographical account of a journey from Gainsborough to Mablethorpe. In your description pay particular attention to the geology, scenery, agriculture and other activities in the districts through which you would pass.

(2) What are the mean January and mean July temperatures, and the total

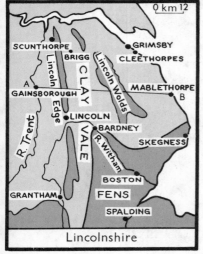

Lincolnshire

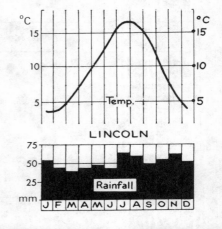

LINCOLN

CHALK ESCARPMENT
Patches of boulder clay
Sheep folding
Arable farming ⟨ roots / cereals

COASTAL PLAIN
Silts and clays
Resort towns
Pasture on drained marshes

CLAY · CHALK · ALLUVIUM · SEA · B

annual rainfall at Lincoln? (3) Compare these figures with those for Rhayader (page 91) and explain the differences. (4) Read page 55 again and explain why summers are wetter than winters at Lincoln.

(5) Explain why: (a) there are sugar factories at Spalding, Bardney and Brigg; (b) arable as well as pastoral farming is possible on the chalk and lime-stone escarpments in Lincoln-shire; (c) many Yorkshire people go to Skegness, Mable-thorpe and Cleethorpes for their summer holidays; (d) agri-cultural implements and ex-cavating appliances are import-ant manufactures in Lincoln; (e) Lincoln is called a 'gap' town.

(6) Can you explain the posi-tion of the villages near Lincoln and the meaning of the network of small channels near Bardney? (*If not, read again pages 30–31.*)

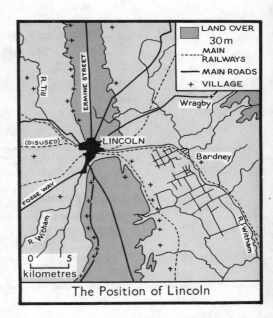

The Position of Lincoln

GAIN . . . This huge mechanical grab is removing 30 metres of soft clay to reach iron ore deposits in the Jurassic escarpment near Scunthorpe.

These ores are worked 'open-cast' at several places along the escarpment between the Humber and Northampton. They account for about 98% of all iron ore produced in Britain to-day, and over half of this comes from quarries at Frodingham, Scunthorpe and Grantham in Lincolnshire.

The output of 'blackband' ore (pp. 79, 94, 124, 146) is now negligible. High-grade haematite (p. 106) is still mined, but in small and decreasing quantities. (7) From the diagram, what percentage of iron ore used in Britain today is imported?

Formerly the removal of Jurassic iron ore left giant scars on the landscape in Lincolnshire, Northamptonshire and Oxfordshire. When the grabs finished work the land was completely derelict: . . . "the sheep have gone and so has the grass on which they grazed—great mounds of earth and rubble rear up in monstrous ridges, and the scene presents all the dismal chaos and desolation

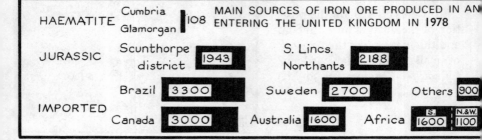

HAEMATITE	Cumbria Glamorgan	108	MAIN SOURCES OF IRON ORE PRODUCED IN AND ENTERING THE UNITED KINGDOM IN 1978		
JURASSIC	Scunthorpe district	1943	S. Lincs. Northants	2188	
IMPORTED	Brazil	3300	Sweden	2700	Others 900
	Canada	3000	Australia 1600	Africa S 1600 N.&W. 1100	

of a battlefield ". Since 1951, however, when Parliament passed a special Mineral Workings Act, great efforts have been made to restore excavated areas. Usually it is possible to replace the top-soil and the land is regained for farming: elsewhere forests are planted. In a small country like Britain there are many competing uses for land—housing, factory building, airfields, power-stations, as well as farming—and the wasteful methods of the past are no longer acceptable.

Land restoration problems do not arise with the most recent excavations, where deeper beds of ore are tapped by means of underground shafts. (8) Explain why these newer mines lie to the *east* of the open-cast district. (See diagram, page 156.)

At Frodingham and Scunthorpe three huge works produce one-eighth of all British crude steel and one-sixth of blast furnace output. The ore from the Lincoln Heights is ' lean ', containing only 20–22% iron. (9) Why is it therefore cheaper to smelt it on the spot, with coking coal brought 25 kilometres from Nottingham-shire collieries, than to send the ore to the coalfield?

Coal is coked, ore is smelted and molten iron converted into steel, all in the same **integrated works**. Time and money is saved by this arrangement, for the molten iron coming direct from blast-furnace to steel-mill needs little re-heating. Imming-ham is a convenient port for importing iron ore and for the export of finished steel goods. (See also pages 119 and 141.)

. . . R E G A I N . Landscaped terrain, including newly planted trees and a golf course on old iron workings, Northamptonshire.

This car is driving across land in south Lincolnshire which three centuries ago lay below 5 metres of water. It would become submerged again if there were some mishap to the protecting dyke or to the elaborate network of drainage canals and pumps which carry off surplus water to the sea. The low-lying district of reclaimed marshland bordering the Wash is called the Fens. (*10*) What other counties besides Lincolnshire are partly in the Fens? (*11*) What rivers cross the area?

The shallow Wash was once much larger, but it has gradually been filled with alluvium brought by rivers and sea currents. The alluvium hindered the rivers' course to the sea, so that lagoons and marshes were formed. As the marsh vegetation decayed deposits of peat slowly accumulated. There were attempts to drain the Fens as early as Roman times, but large-scale reclamation did not begin until the 17th century. Until then it remained for the most part a remote, desolate marshland, where a few scattered inhabitants gained a livelihood by fishing and hunting wildfowl. They lived on small islands in the Fens where patches of gravel or boulder clay rose above the flood waters.

The 4th Earl of Bedford played a leading role in the first really effective drainage scheme. Between 1630 and 1651 be brought in Dutch engineers who straightened the course of the Bedfordshire Ouse by means of two parallel cuts called the Old and New Bedford Levels. A network of tributary ditches carried water to the main channels. At first all went well: the water-level fell, so that pastures and cornfields soon appeared where formerly had been marsh and bog. However, the peat shrank as it dried and the level of the reclaimed land gradually

Fenland

fell—in places by more than 4 metres. The carefully designed drainage channels were disrupted and floods returned. As many of the drainage ditches now stood higher than the surrounding Fens it was necessary to install pumps to get rid of the flood waters. At first these were hand-operated or driven by windmills —today they depend on electric or diesel power.

The flat Fenland scenery, with its windmills, dykes and bulb fields, is very like that of the Netherlands. Many villages and market towns still cling to the isolated ' islands ' of slightly higher ground. " I travelled towards Ely in the early morning long before the first harvester was awake. At this time of the year a veil of white mist lies over the Cambridgeshire fenlands, a pearl pale thing, thin and chill; and as I went on through it I felt as though I were sailing on the ghost of a sea. The dimly seen hedges of this flat chessboard land were like the edges of poised breakers. Suddenly I saw before me, like a frozen ship upon a frozen ocean, the Isle of Ely rising in spectral beauty above the morning mist. This sudden high hill crowned with its towered cathedral seen above the white mist of late summer is one of the most beautiful things in the whole of England. It is a spellbound hill: the creation, it seems, of a wizard's wand: . . . As the sun rises and the mists melt, the Isle of Ely—the Isle of Eels is the real name—grows to reality, becomes a little town clustered round its old cathedral; but even in the full sunlight it never quite loses its air of having been built by magic." [1]

FARMING IN THE FENS

The reclaimed Fens form one of Britain's most valuable agricultural districts. The peat fen is particularly fertile, especially when mixed with the underlying chalky clays. Addition of clay also reduces the risk of wind-erosion of the light peaty soils. Cultivation is intensive, and there are many market gardens. Most of the land is ploughed. The chief crops are:—

POTATOES	The Holland district of Lincs. and the Isle of Ely is Britain's largest potato-growing district. Most go to London markets.
SUGAR-BEET	Factories at Ely, King's Lynn, Peterborough and Spalding.
FRUIT	Especially around Wisbech, where there are canning factories.
BULBS	Spalding and Wisbech bulb-growers are world-famous.
WHEAT, oats, and barley	Climate in the Fens particularly favours cereals, for there is both rain and much sunshine in summer. ((12) Why?)

[1] H. V. Morton, *In Search of England*, op. cit.

This large boulder at East Bergholt, in north Essex, has lain 'stranded', at least 80 km from the nearest beds of hard rock, since long before human beings settled in this area. It is too heavy for any flood to have washed it so far. How, then, did it get here? The answer is suggested in the diagram below.

CHAPTER 13

East Anglia

ON PAGE 32 we saw that a huge ice-sheet once lay over all Britain to the north of a line joining the Severn and Thames estuaries. East Anglia (Norfolk, Suffolk and north Essex) was on the fringe of this sheet and the land there was much affected by the ice and its melt-waters. Extensive deposits of glacial clays, gravels and sands are spread over the region, together with 'erratics' like that illustrated above.

The great Scandinavian Ice Sheet was stationary for some time over the northern part of Norfolk and a huge terminal moraine accumulated there. In places reaching over 100 m high, it can be traced for 25 km across the countryside near Cromer. Meltwaters pouring southwards spread patches of sands and pebbles from this moraine over large portions of Norfolk and Suffolk.

Much fine dust (see page 36) was also washed out from the

Diagrammatic Section across the East Anglian Heights in Glacial Times

NW

GREENSAND   CHALK   LONDON C

As the ice gradually pushed its way southwards it came up against the hills of the Chalk outcrop ((1) *known by what names in East Anglia?*) Its enormous force enabled the moving ice to push over the crest of these hills, breaking off masses of chalk on its way. Crushed and mixed with clay and rock fragments, this chalk was dumped on the dip-slope to form a deposit known as **Chalky Boulder Clay**. (2) Illus-

trate this by copying the diagram abov and adding the following labels:

ICE SHEET / CHALK OUTCROP / CHALKY BOULDER CLAY / FORMER CREST-LINE

Show by arrows the direction in whic the ice is moving. (3) Why do you thin the stratum lying above the Chalk is calle London Clay? (*Hint: cross-section p. 169.*)

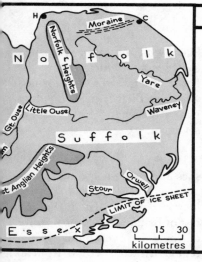

At Hunstanton (H) the Chalk escarpment forms prominent cliffs, but the Chalk dips gently eastwards until at Cromer (C) it lies below sea level. ((4) *Compare this map with the diagram on page 29. Which of the main rivers of East Anglia are ' dip-slope streams '?*)

Thus, in the western districts of East Anglia, the Chalk lies at or near the surface and forms a rolling countryside. The east is a generally level lowland covered by later deposits. These are mainly of glacial sands, gravels and boulder clay.

Chalk is quarried in many places for cement and for agricultural lime. Sand and gravel quarries are equally common, especially in the east. (5) For what purposes are sand and gravel chiefly used?

Cromer moraine. Mixed with sand and boulder clay, it now forms extremely fertile loamy soils in parts of north-east Norfolk. (Loams retain plant foods and moisture and are easily penetrated by roots.) Notice in addition:—

(*a*) The Breckland: an area of glacial sands, gravels and sandy boulder clay between the East Anglian and Norfolk Heights. Both rainfall and soils are very light.

(*b*) The Broads: a district of partly reclaimed marshland dotted with stretches of open, shallow water. At one time there were tidal estuaries here like those farther south on the Suffolk and Essex coast. Longshore drift (as described on page 39) sealed off the inlets from the sea to form lagoons which were gradually

For long the Breckland remained unproductive: now, in the past fifty years, the largest coniferous forest in Britain has been planted there by the Forestry Commission. Mature trees are already being felled for use in local factories and elsewhere.

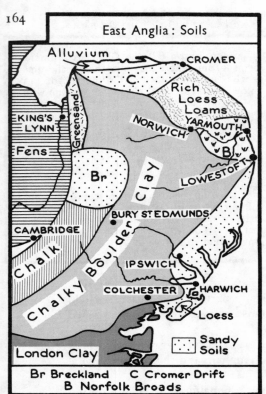

East Anglia : Soils

Alluvium
CROMER
C
Rich
Loess
Loams
KING'S
LYNN
NORWICH
YARMOUTH
Fens
Greensand
B
Br
Clay
LOWESTOFT
BURY ST. EDMUNDS
CAMBRIDGE
Chalk
Chalky Boulder
IPSWICH
COLCHESTER
HARWICH
London Clay
Loess
Sandy Soils

Br Breckland   C Cromer Drift
B Norfolk Broads

filled with alluvium deposited by rivers. Much later the Broads themselves were formed when medieval peat diggings were 'drowned' by a rise in sea level.

Protected by Fenland marshes and dense oakwood forests on heavy clay, East Anglia for long remained isolated from the rest of Britain. Yet the fertility of the soils over much of the region attracted settlers from an early date. The Anglo-Saxons established a kingdom here and it was they who first tackled the thick forests. (*(6) The Saxons came by boat—from where? Do you think their landings were helped or hindered by the coastline in this part of Britain?*)

In the late 18th century new methods of arable farming made cultivation possible even on the dry, less fertile chalk downs and sandy districts in the west of the region. These had formerly been used only for sheep rearing, but now turnips were grown and the sheep 'folded' on the turnip fields. The sheep ate the turnips and their manure enriched the soil, which was later sown with barley, wheat or clover. Turnips and clover were also fed to cattle. Well-trodden with manure, the straw used for their bedding went back on the land every autumn. As each field was used in turn, what 'corn' took out of the soil 'horn' put back. So began the famous Norfolk four-course **rotation of crops*** which in time transformed British farming by building up the fertility of poor soils. (7) Who were 'Coke of Norfolk' and 'Turnip Townshend'?

Today, though heath and woodland still cover the poorest soils, the broad fields of East Anglia are among Britain's most valuable

*See Book 1, ch. 3, where an East Anglian farm is studied in detail.

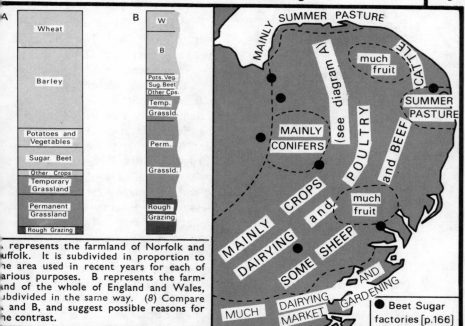

A represents the farmland of Norfolk and Suffolk. It is subdivided in proportion to the area used in recent years for each of various purposes. B represents the farmland of the whole of England and Wales, subdivided in the same way. (8) Compare A and B, and suggest possible reasons for the contrast.

● Beet Sugar factories [p.166]

farmlands. Cereals are the leading crops: barley mostly on the lighter, sandier soils, and wheat on the heavier loams and clays.

Sugar beet is next in importance. Apart from its value as a cash crop and as a 'break' between two cereal crops, the tops and pulp make excellent stock feed. In recent years carrots, peas and other vegetables, grown under contract to canning or freezing firms, have become a popular sideline on large arable farms.

Turnips and swedes have almost disappeared, because sheep and even cattle have dwindled in numbers wherever cereals can be grown. The four-course rotation is no longer used in its simple form. Chemical fertilisers have largely replaced 'muck', and machines have largely replaced men. Scenes like that on page 243 are common throughout west Norfolk and Suffolk. Farms of 200–500 ha contrast sharply with the 15–30 ha typical of the nearby Fens. (9) Reread page 161 and explain this contrast.

(10) Using the maps and diagrams, write an account of present-day farming in East Anglia. Explain carefully the influence of (a) relief (b) climate (c) surface geology (d) new ideas.

(11) Suggest a large nearby market for the dairying and market gardening of north Essex.

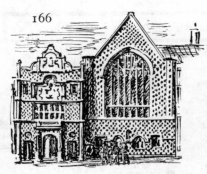

The Trinity Guildhall at King's Lynn on the west Norfolk coast has a somewhat Dutch look about it. Like many other similar buildings in East Anglia, it is a reminder of the close trading links which this part of Britain once had with Flanders. In early medieval times large quantities of wool were sent from East Anglian ports for weaving in **Flemish towns like Ghent and Bruges.** Then East Anglia itself became one of the thriving centres of the English woollen industry. The wool merchants were rich, and the magnificence of many East Anglian churches bears witness to their generosity. Norwich was at one time the leading manufacturing city in Britain. The village of Worstead, near Norwich, gave its name to the worsted cloths for which East Anglia was famous.

Then came the Industrial Revolution, and with it the rapid decay of East Anglian woollen manufacturing. A few silk and rayon mills are all that remain today of textile working there. (*12*) What did East Anglia lack that gave other woollen centres like West Yorkshire an overwhelming advantage in the new era?

Having a surplus of skilled but unemployed textile workers to hand, enterprising Norwich businessmen started the manufacture of women's footwear; this is still among the city's main industries. Throughout the 19th century Norwich kept its place as a social and commercial centre for the region. With its cathedral, university, livestock market and shopping centres, Norwich (*about 140 000*) is very much the 'capital of East Anglia', though Ipswich (about the same size) and Colchester are important local centres and have similar industries.

(*13*) From this list, which of Norwich's industries (*a*) reflect its position as a regional centre? (*b*) have some link with farming? Explain your answers.

(*14*) Suggest (*a*) why half of Britain's beet sugar factories are in East Anglia and (*b*) why those factories (*map p. 165*) are so widely scattered. (*Hint: only about 13 tonnes of sugar can be extracted from every 100 tonnes of sugar beet.*)

All round the coast, from King's Lynn to Southend, there are towns and villages which between them cater

**TYPICAL
PRODUCTS
AND
INDUSTRIES
OF NORWICH**

Electrical engng.
Printing
Insurance
Mustard
Flour
Footwear
Banking
Beer and vinegar
Farm machinery
Furniture
Chocolate
Poultry sheds
Wire netting
Clothing
Fertilisers
Sacks
Animal feeding stuffs

(15) Use your atlas to identify the ports marked on the map and referred to below.

The trade links that bound East Anglia [to] Flanders in the Middle Ages grew up [be]cause the English coast is here so close [to] the Continent. Similar links remain to [thi]s day, for H is the prime ferry-port for [Be]lgium, the Netherlands and Germany. [F] was developed in the 1960s as a 'roll-on, [rol]l-off' and container port. Its trade is [no]w greater than that of H. The ports of the Stour-Orwell estuary (F, H, I) together handle a greater value of trade than any other British group except the huge Port of London. As well as mail and passenger traffic, there are large imports of perishable foods (e.g. meat, eggs, butter, fruit), paper, machinery and vehicles. The exports are mainly machinery and vehicles.

L, Y, K are minor ports but their trade, too, has greatly increased as Britain's links with Europe have become closer.

for holiday-makers of all types. This has become one of the region's major industries during East Anglia's short but sunny and dry summer. The inland towns are mostly small market centres with a little light industry. Some, such as Haverhill, Thetford and Bury St. Edmunds, have expanded to take the 'overspill' of people and light industry from the London area. King's Lynn, too, is an 'overspill' town. Canned and frozen vegetables are among its main products. (16) Why?

The famous herring fisheries on which Yarmouth and Lowestoft once depended are dead, killed probably by over-fishing of the North Sea breeding grounds. Lowestoft is still a fairly important trawler port, but Yarmouth now lives largely on its holiday trade ('The Blackpool of the East Coast'), on food processing (potato crisps, frozen peas, etc.) and on engineering and other services required by the North Sea gas and oil rig crews.

(17) Suggest why heavy industry has never become established in East Anglia.

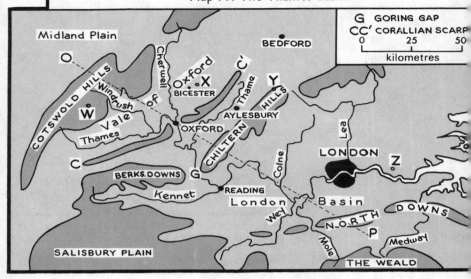

Map A: The Thames Basin

CHAPTER 14

The Thames Basin

A **river basin** is the total area drained by a river and its tributaries. Headstreams draining to different river basins are separated by higher land called the **watershed, water parting,** or **divide.** A watershed is thus a zone around the edge of a river basin, separating it from neighbouring basins.

The largest river basin in Britain is that of the Thames. On Map *B* the River Thames is shown, with its tributaries and adjacent rivers. (*1*) Make a tracing of the map, name the streams, **and continue the line** *AB* to show the Thames Basin watershed. (Your line should return to the Thames Estuary at *C*.)

(*2*) Now look carefully at the diagram and maps and sort out the following ' heads and tails ':—

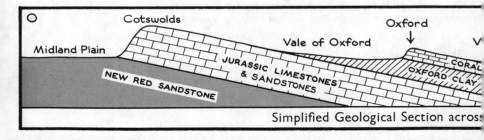

Simplified Geological Section acros...

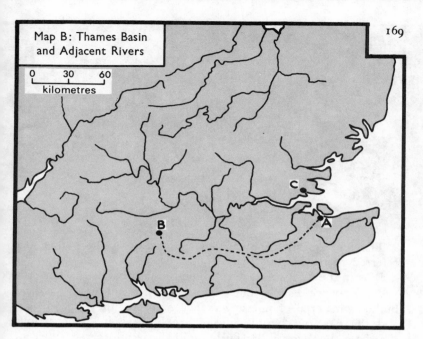

The source of the Thames is	*up the Kennet valley*
The Corallian escarpment is	*on the dip slope of the Cotswolds*
The Chiltern's scarp slope faces	**?** *pierced by the Thames at Goring Gap*
The Chalk ridge is	*across the Vale of Aylesbury*
The route west from Reading runs	*broken by the Thames at Oxford*

Next see how many of the following you can answer correctly:—

(3) Some of the strata in the diagram below are in the form of a syncline—which are they?

(4) What do you think was the origin of the Corallian Limestone?

(5) Of the London and Oxford Clays which is the newer sediment? (Give reasons.)

(6) What other main river basins lie adjacent to the Thames Basin?

(7) What rock would you expect to find if you dug away the surface soil at the places marked *W X Y Z* on Map A?

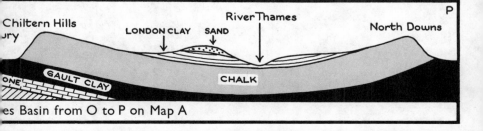

es Basin from O to P on Map A

W

The four quarries shown here are located at the points just referred to on Map *A* (page 168).

W is in the Cotswolds, where limestone is obtained. The pale yellow Cotswold limestone (Bath stone) is an easily worked building material of fine appearance—most of the local villages and the near-by Oxford colleges are built of it. One serious drawback, however, is its fairly rapid decay under the attack of acidic rainwater. (Look again at page 19.) Near Banbury in the Cotswolds the limestone is quarried for its iron-ore content (see page 158).

X is near Bicester, in the Oxford Clay Vale. The photograph below shows one of the London Brick Company's huge excavations. Notice in the picture that excavators are stripping off an ' overburden ' of some 5 metres of boulder clay which lies above the Oxford Clay at this point. More than thirty large-scale brickworks are concentrated in the Clay Vale between Bicester and Peterborough. After excavation the clay is taken to the brickworks adjoining the quarry, where the bricks are pressed into shape mechanically, and baked.

X

Y is at Dunstable at the foot of the Chiltern escarpment. **Y**
Here, from a quarry a kilometre long, chalk is excavated for manufacturing cement in the adjacent factory. Similar cement works
using Chilterns chalk are found near Luton and Tring. There
are also enormous chalk quarries for cement-making along the
banks of the lower Thames in Kent and Essex. ((*8*) *What chalk
ridges are worked here?*) The chalk from the Chilterns is sufficiently clayey for direct manufacture into cement. Clay has to be
added to the lower Thames chalk, however, and is quarried from
the near-by London Clay. The works beside the Thames can
easily import coal and despatch cement by boat.

Z is a gravel pit in south Essex. There are many other similar
pits on the sand and gravel terraces which line both banks of the
lower Thames. About one-third of Britain's gravel output comes
from this source—the rest being produced from glacial deposits
(see page 163) and 'solid' deposits such as the Bunter Pebble
Beds in the Midlands (see page 31). There are over 1000 gravel
pits in the country, and the total tonnage output annually ranks
next to that of coal-mining. Sand and gravel is chiefly needed to
make concrete. This photograph shows a 'wet' pit, i.e. one in
which material is being scooped from below the water-table.

The scars left on the landscape, and the land-use problems
involved, are like those of open-cast iron-mining (see page 158).

Z

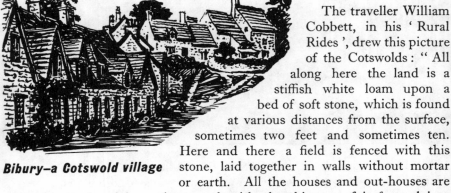

Bibury—a Cotswold village

The traveller William Cobbett, in his ' Rural Rides ', drew this picture of the Cotswolds : " All along here the land is a stiffish white loam upon a bed of soft stone, which is found at various distances from the surface, sometimes two feet and sometimes ten. Here and there a field is fenced with this stone, laid together in walls without mortar or earth. All the houses and out-houses are made of it, and covered with the thinnest of it formed into tiles. The stiles in the field are made of large flags of this stone, and the gaps in the hedges are stopped with them. Anything quite so cheerless as this I do not recollect to have seen; these stones are quite abominable." This description, written early in the 19th century, is still accurate for many parts of the limestone escarpment, but few people will agree with Cobbett's last remark—Cotswold towns and villages are generally considered to be amongst the most picturesque in Britain. The warm mellow shade of the weathered stone blends perfectly with the tawny loam soils.

The Cotswolds have long been famous for sheep-rearing, and were once an important centre of woollen cloth manufacture. As a result of the Industrial Revolution the cloth industry here suffered the same fate as that in East Anglia (see page 166). The few remaining factories make specialised products of high quality. (9) How has this helped them to survive? The chief centres are Stroud (military dress uniforms, hunting costumes, billiard table cloth), Chipping Norton (whipcords) and Witney (blankets). The picture below shows the medieval wool merchants' market hall at Chipping Campden.

Although there are still many sheep on the higher, more barren parts of the Cotswolds (which in places reach 300 m), very few open downs remain. In the valleys of the Thames head-streams, and on the more fertile parts of the dip-slope, mixed arable farming and dairying predominates.

This old cottage stands in a village near Oxford. The brick and timber of which it is built is evidence of local clay and former oakwood forest. The damp clay soils in the Vale of Oxford have been farmed since Saxon times, when the first inroads were made on the dense woodlands. Ploughing is difficult on the heavy clay, and much of the land is left as permanent pasture for large dairy herds. Although the Oxford region is still fairly rural it is undergoing a rapid increase in population and employment. For many years the ancient university city of Oxford has had an industrial quarter at Cowley, with large factories at British Leyland and Pressed Steel. Now other towns in the Vale of Oxford have fast-growing new light industries. Both Swindon and Aylesbury, for example, have firms engaged in mechanical and electrical engineering, food processing, printing and publishing. Similar developments are transforming Banbury and Bletchley, and in Witney more people are now employed making machine tools and motor accessories than the more famous blankets. There is also a remarkable number of Government and industrial research establishments around Didcot.

Major attractions of this region for new firms include (*i*) plenty of land on which to build and expand and (*ii*) good fast road and rail links to nearby London and Birmingham. Completion of the M4 motorway has also given the region better access to ((*10*) *which?*) ports. (Map, page 281.)

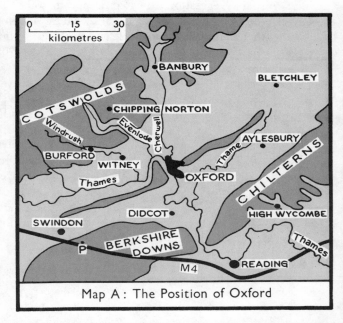

Map A : The Position of Oxford

The chalk escarpment of the Berkshire D
and Chiltern Hills rises sharply from
southern edge of the Vale of Oxford.
can obtain fine views like this all along
scarp slope from Swindon to Luton.
photo was taken from point P on the m
the previous page. Notice the striking
trast between the cultivated clay plain,
the slopes of the escarpment where the g
ent is too steep for ploughing.

Such steep parts, together with districts where the chalk soils
are too thin for cultivation, form true ' downland '. It is covered
with short, springy turf, and makes good sheep pasture.

Very little open downland remains today. Although chalk is
porous, the rainfall (750–1000 mm) on the escarpment is sufficient
to permit large-scale arable cultivation where there is sufficient
depth of soil. Huge tracts of chalk downland in south-east
Britain were first cultivated during the Second World War, with
the aid of tractor-drawn ploughs. Wheat and barley are the chief
crops.

Many people are surprised, on first visiting the Chilterns, to
find extensive woodlands of elm, beech and oak, instead of the
sparse vegetation expected on porous soil. The woods are found,
too, mostly on higher ground, where one might reasonably think
the chalk would contain least moisture. The reason for the presence of the trees is a thin
layer of a deposit called
' **clay-with-flints** '. This
sticky, water-holding clay is
a residue of impurities from
within the chalk which, now
dissolved away, formerly
lay above the present land
surface.

The typical rounded chalk
hills are thus often crowned,
like the one in this photo-

LAND OVER 150ᵐ

HITCHIN

AYLESBURY   LUTON

TRING

OXFORD   WENDOVER

PRINCES
RISBOROUGH

DIDCOT   HIGH
WYCOMBE

GORING

READING   LONDON

0   10   20
kilometres

Railway Routes through Chiltern Gaps

graph, by clumps of woodland. Where it has been cleared the 'clay-with-flints' supports cattle, on damp pastures. These contrast remarkably with the adjacent dry chalk slopes.

Soils differ again on the floors of the deeper valleys which pierce the chalk ridge. Here alluvium and gravel deposits allow both cattle rearing and arable farming. In the valleys, too, are the principal towns. They grew up as **' gap ' towns** where roads, canals and railways converge to cross the escarpment. (Compare Lincoln, page 157). Many of the valleys have small woodworking industries which originally used timber, especially beech, from the surrounding hills. The largest firms are in High Wycombe, where there are more than a hundred concerns making chairs and furniture. Other factories produce paint, varnish and furniture-making tools.

(*11*) Now read pages 30–31 again and then on a large copy of the diagram below insert the following labels in their appropriate places:—

SATURATED CHALK,   CLAY-WITH-FLINTS,   DRY VALLEY, GAP-TOWN,   UNSATURATED CHALK,   DRY CHALK SOIL, WATER-GAP,   GRAVEL AND ALLUVIUM,   WATER-TABLE, SHEEP,   CATTLE,   ARABLE.

(*12*) Make a large copy of the map opposite and add labels to show the destinations of the railway routes. (Use an atlas.)

Diagrammatic Section through Chalk Country

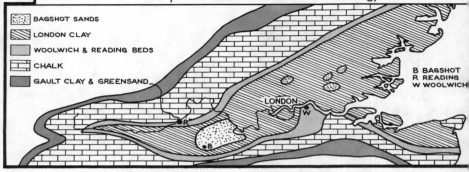

Map A :— The London Basin : Geology

BAGSHOT SANDS
LONDON CLAY
WOOLWICH & READING BEDS
CHALK
GAULT CLAY & GREENSAND

The London Basin is a V-shaped syncline. (See p. 12.) The angle of the V rests in the Vale of Kennet, and the open end is partly submerged by the sea. The syncline is bounded to the north by the Chilterns and East Anglian Heights, and to the south by the North Downs. This higher rim of the Basin often exceeds 250 m: it would be a serious hindrance to communications were it not for the many gaps, such as those cut by the Thames at Goring and the Medway at Chatham. (*Maps, pp. 174 and 177.*)

As the London Basin lies in south-east Britain, the equable influence of the Westerlies is not very marked. Rainfall is less, and the temperature range greater than in most other parts of Britain. (*13*) Use the table opposite to make a large labelled sketch-map to show land use in the London Basin.

No part of the London Basin is more than two hours travelling time from central London, and virtually every settlement has its

Geological Section across the London Basin from X to Y on Map B

Structurally the Basin is a syncline of chalk which at one time contained an inland sea. Sand, gravel and clay sediments which were deposited in this sea now form the rocks within the Basin. To these, in recent geological times, have been added glacial deposits and river alluvium. On page 31 we saw that the London Basin is a source of artesian water.

(*14*) Make a large copy of the diagram below and on it add the following labels their correct places (the maps will help):

VALE OF AYLESBURY
VALE OF HOLMESDALE
VALE OF ST. ALBANS
CHILTERN HILLS   NORTH DOWN
HAMPSTEAD HEIGHTS   THAM
WOOLWICH and READING BEDS.

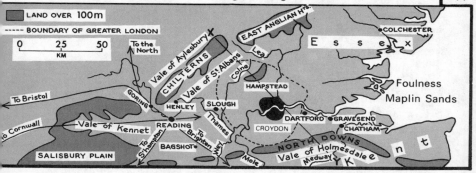

quota of commuters.   Almost 3 million people go to and from
work in London each day, the great majority of them travelling
most of their journey by suburban train and tube.   Many British
Rail stations provide large car parks for the convenience of commuters, and former rural communities throughout the Basin have
been changed into **dormitory suburbs.**   (*(15) Explain this term.*)
Many commuters live on the outer fringe of Greater London,
where new housing and industrial estates were built at a time when
the London County Council was encouraging employers to move
out of the capital's centre.   Typical of such locations is Croydon
(330 000), where new light industries and fifty office blocks
provide work for 30 000, but which is also the greatest commuter
centre south of the Thames.

A potential industrial growth centre is Maplin (*see map*), a
location with many advantages for the construction of a super-
tanker and bulk-carrier terminal.

DISTRICTS OF THE LONDON BASIN

Vale of St. Albans.   Light soils favour market gardening.   Much glasshouse produce,
e.g. tomatoes, cucumbers, flowers, early vegetables.   High prices fetched in nearby
London markets.   Dairy farming and pig-rearing also important.

Essex.   Much milk produced for London.   One of Britain's driest counties (rainfall
c. 625 mm), but heavy London Clay soils retain moisture.   Good grass covering throughout year.   Market gardening on better-drained boulder clay soils.   Many oyster beds
on mudflats of low, marshy coast, especially near Colchester.

Middle Thames.   Many small riverside towns, e.g. Henley, Marlow, Maidenhead
provide summer ' playgrounds ' for Londoners.   Slough Trading Estate has good road
and rail transport and huge market in nearby London: light industries include electrical
apparatus, patent foods, radio equipment, vehicles.

Vale of Kennet.   Military training grounds on infertile sands at Bagshot.   Vale contains main road and rail routes to the west.   Farming on chalklands like that of Chilterns (see pp. 174–175).   Reading: market town and route centre (see map); industries
include engineering, biscuit making and seed growing and packing.

North Kent and South Essex ('Thamesside').   Many industries which require abundant
space, or use bulky materials like iron ore, suited to water transport.   Examples
include: cement works at Northfleet and Purfleet; paper works at Gravesend, Dartford
and Purfleet; petrol refining at Shellhaven, Purfleet and Coryton; vehicle manufacture
at Dagenham; vegetable oil refining and margarine production at Purfleet.

CHAPTER 15
London

WE THINK of London as a vast, roaring metropolis—a sprawling city with 7·6 million inhabitants. It is difficult to imagine the desolate riverside encampment which marked man's early settlement on the north bank of the Thames.

London first became an important town in Roman times, as a link between the two most settled and civilised parts of Roman Britain—Essex and Kent. Two of the earliest Roman forts were built at Verulamium (St. Albans) and Camulodunum (Colchester), and the Thames was a barrier between these centres and the Channel ports. The main route into Britain was Watling Street, which followed the North Downs to the south bank of the Thames at Greenwich. To cross the river here (*G* on map) would have been very difficult—the map shows why. A little farther west, ridges of comparatively dry gravel reach right down to the river's edge on both banks. At first there was a fording place (*FF'*) where the river flowed from south to north with a broad, shallow channel. A low sandy island (*W*) at Westminster made the crossing easier. To guard the ford the Romans built a fort on a near-by gravel mound (*H*1). It was surrounded by dense forests, marshes and lagoons, but was a good defensive site. A bridge which was built here was probably reached from the south wooden causeway (*C*) over the alluvial marsh.

The first Roman fort was destroyed in A.D. 61 by rebellious Celts. When the Romans rebuilt the town they extended the camp to include another gravel mound (*H*2), and enclosed the new settlement with a wall. This ancient walled fortress is the site of the present City of London (see map, page 187). The gravel mounds are today crowned by St. Paul's Cathedral and Leadenhall Market. Two small north-bank streams, the Fleet and Wallbrook, are now built over, but the former is marked by famous Fleet Street. In time the bridge at London became the focus of six main Roman roads and a trading port grew up with links throughout the Empire. The Roman bridge had such low arches, however, that large vessels could not pass upstream and thus London became a transhipment point where

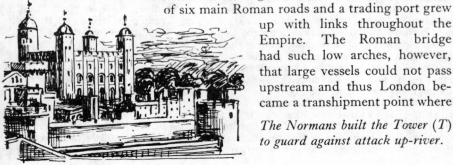

The Normans built the Tower (T) to guard against attack up-river.

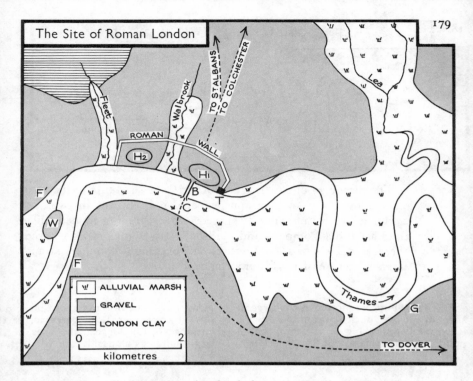

The Site of Roman London

ALLUVIAL MARSH
GRAVEL
LONDON CLAY

0   kilometres   2

cargoes were discharged and reloaded on to smaller craft. Its importance as a bridging point, route centre and trading place has increased through the centuries.

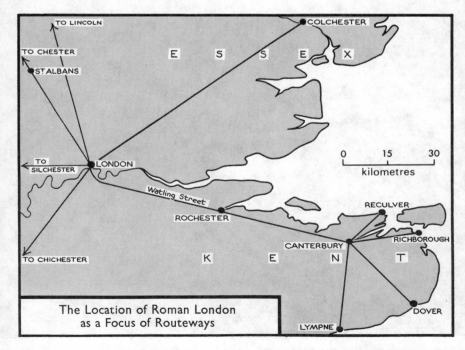

The Location of Roman London
as a Focus of Routeways

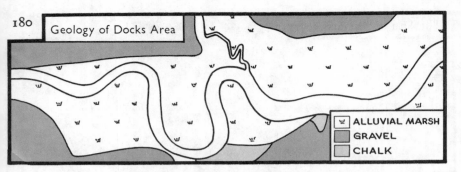

Geology of Docks Area

ALLUVIAL MARSH
GRAVEL
CHALK

The discovery of America and of the Cape sea-route to the East was followed by an enormous expansion in European overseas trade, in which London fully shared. Great trading concerns like the East India Company and Hudson's Bay Company were based there, and by A.D. 1600 London had become a commercial and financial centre of world importance.

This aerial photograph shows the largest sheet of enclosed dock water in the world. It was taken looking west from above point X on the map opposite. (1) How many dock basins can you pick out (2) Why are there lock gates across th entrances? (3) Suggest the uses of th large buildings alongside the wharves.

Map A : The London Docks

City — LONDON DOCKS — E. INDIA DKS. — W. INDIA DKS. — ROYAL VICTORIA DOCK — ROYAL ALBERT DOCK — Beckton

Tower Bridge — SURREY COMMERCIAL DOCKS — MILLWALL DOCKS — Transporter Yard — KING GEORGE V DK. — River Thames → — Woolwich — Greenwich

Lea

0   2   4 kilometres

* Docks closed

The upstream location of the London Docks and the Surrey Commercial Docks made them less accessible to large modern cargo vessels, and both docks closed in the late 1960s. Together with other disused dockland at Millwall and Beckton the Port of London Authority thus has 338 hectares available for redevelopment. All future extensions will be well down-river at deep 'tidewater' sites and so the derelict dock areas provide planners with '. . . the greatest opportunity in London since the Great Fire.' (4) Suggest, with reasons, the most suitable uses for this land. Tilbury (map, p. 185), which has long been London's main passenger terminal, has large new container docks. Suggestions for the future include a large cargo port at Maplin Sands (map, p. 177).

The Port of London developed downstream from the **head of navigation** at London Bridge. The map above shows the system of docks which were built between 1802 and 1921. (5) How did the geology of this part of the Thames Basin favour their excavation?

The various main docks deal with particular types of cargo for which they are equipped with special handling gear and storage sheds. Thus huge quantities of bananas and meat are landed in the Royal Albert Dock and of grain in the Millwall Dock.

About one-third of the enormous variety of cargoes brought to London docks is loaded into lighters like those in the scene below. The lighters are constantly plying between the docks, quays and factories on the river bank, as well as reloading goods on to small coastal vessels for reshipment to other British ports or the Continent, for the Port of London is a great **entrepôt**. London is also Britain's largest port, handling the percentages shown below of the U.K.'s imports of certain commodities. However, the

Oil seeds/nuts	66	Vegetable oils/fats	21
Sugar	37	Rubber	17
Paper	36	Fruit and vegetables	14
Woodpulp	34	Dairy products and eggs	12
Grain	30	Meat	11
Wood	24	Chemicals	9

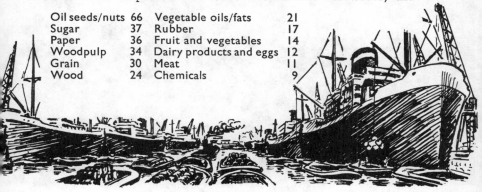

PORT OF LONDON: VOLUME OF TRADE ('000 tonnes)								
	1970	1971	1972	1973	1974	1975	1976	197.
Upper Docks	4988	4589	3461	3069	2596	1679	1810	178
Tilbury	5324	5767	6116	6329	6140	5142	6420	672

PORT OF LONDON: SHARE OF U.K. PORTS TRAFFIC BY VALUE (%)								
	1970	1971	1972	1973	1974	1975	1976	197.
Imports	19.1	19.1	17.3	15.1	12.9	10.9	11.5	10.8
Exports	25.1	23.1	21.6	20.0	15.6	14.0	12.7	11.8

total volume of trade handled by the Port of London, and the Port's share of total U.K. ports traffic, have both dwindled in recent years. (6) Draw charts to illustrate the information given in the above table.

This decline in trade has helped to bring about **inner city decay** in the dockland boroughs of Newham, Tower Hamlets, Southwark, Lewisham and Greenwich. (*See map, p. 184.*) Decay has also resulted from the movement of many local firms—encouraged until recently by the policy of decentralisation—to 'Expanding' and 'New' Towns in other parts of the country. Over 120 firms, for example, left Newham between 1968 and 1978, whilst Tower Hamlets and Newham lost 17 600 jobs (21% of the total) in the period 1973-76. These firms took with them the bulk of their skilled labour, with the result that two-thirds of the remaining men are manual labourers. Clearly these workers find it difficult to find *any* jobs, not least because the smaller dock cargoes are being handled increasingly by machinery.

The dockland boroughs, with the adjacent boroughs of Hackney, Lambeth and Islington, have been designated a 'Partnership Area'. A three-year area reclamation programme, started in 1978-9, will use Government money to help purchase land, build homes, factories, shops and other services, with the aim of providing up to 12 000 jobs by 1986.

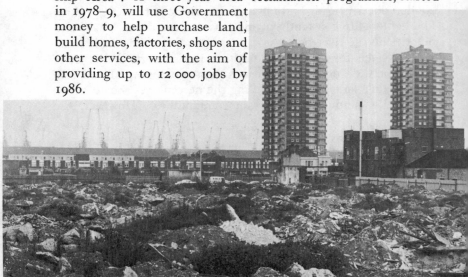

Inner city decay and redevelopment in London's dockland.

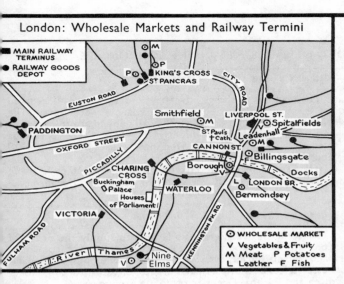

London: Wholesale Markets and Railway Termini

■ MAIN RAILWAY TERMINUS
● RAILWAY GOODS DEPOT

⊙ WHOLESALE MARKET
V Vegetables & Fruit
M Meat   P Potatoes
L Leather   F Fish

Every day a vast amount of produce pours into London from the rest of Britain, as well as from overseas. To handle the raw materials and foodstuffs great wholesale markets have grown up. The most important of these are shown here. Some, like Billingsgate fish market, are extremely old. They grew up beside the Thames when barges were the chief means of transport. Others, like the potato markets at King's Cross, owe their location to the building of railways. All are within easy reach of the ancient City, where banks and merchant houses handling the goods have been established for centuries. The extraordinary importance of London as a world trading centre is symbolised in the group of British and international banks, insurance companies, warehouses, commodity exchanges and business houses between St. Paul's and the London docks.

In some of these markets, like those for tea, sugar and jute, all transactions are dealt with by samples. Such ' markets ' are merely certain streets where City firms which deal in these goods have their offices. However, for perishable goods like fish, fruit, meat and vegetables, there are true markets, to which huge tonnages are sent every day for grading, sale and despatch. Prices paid for these goods in shops all over Britain are thus determined in the London markets.

During the night thousands of lorries and many special trains throng into London with produce for sale. By breakfast time the goods are well on their way to retailers in London and farther afield. Such large-scale movement of produce is possible only because of the many road and rail routes which radiate from the capital. (*See maps, pp. 174 and 281.*) Nearly 3 000 000 people also travel along these routes daily to and from their work in London. Half a million of them flood into the ' square mile ' of the City—which has a resident caretaker population of only 4000.

The majority of London's horde of daily travellers live within 40 kilometres of the City, but some come from as far as Brighton and Ramsgate. Fast electric trains and buses to the suburbs, and the underground railway system, help relieve the traffic congestion.

183

SOME IMPORTANT INDUSTRIES IN THE LONDON AREA

Industries using bulky raw materials	
Cement manufacture Paper manufacture Sugar refining Flour milling Oil refining Electricity generation Chemicals Engineering Metal working	The map below shows how these industries are all situated on the banks of the Thames or navigable waterways. The bulky raw materials on which they all depend are transported most cheaply and conveniently by water. The Battersea electric power-station, shown opposite, uses between 15 000 and 18 000 tonnes of waterborne coal every week. ((7) *Can you name the materials used by the other riverside industries?* (8) *Suggest why oil refining and explosives manufacture are located in such remote locations.*)
Long-established 'East End' industries	
Furniture making Clothing manufacture Boot and shoe making Food, drink and tobacco preparation Cardboard-box making	These occupations grew up largely to supply the needs of London's vast population. Production is mostly from hundreds of small firms, each with a handful of employees. Various districts deal in particular trades, e.g. furniture in Shoreditch and Bethnal Green, clothing in Stepney.
'New' industries, manufacturing articles such as:	
Radio sets Discs and tapes Electrical apparatus Television sets Plastics Cosmetics Confectionery Patent foods	These industries, using electrical power from the Grid (see page 264), have grown up rapidly in recent years in west and north-west London. Scores of factories are strung out along the main rail and road routes out of London. ((9) *Why there?*) Many of them use semi-finished light metals from the Midlands. The huge London market is at hand, and distribution is easy to all parts of the country.

OTHER IMPORTANT OCCUPATIONS IN THE LONDON AREA

'White collar' occupations and service trades (see also page 263)	
Commerce Professions Domestic Transport Dock labouring	e.g. office staff, shop assistants, bank clerks. e.g. doctors, lawyers, dentists, teachers. e.g. hotel and café workers. e.g. bus drivers and conductors, train drivers, porters. (There are over 5000 labourers in London docks.)

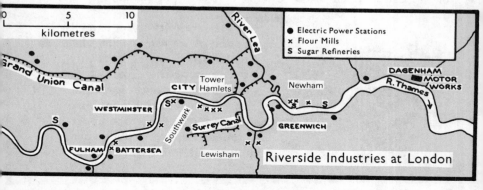

Riverside Industries at London

With 20 000 firms employing a total of 1.2 million workers, Greater London is by far the largest manufacturing region in the country. Details of its remarkable variety of industries are given on the left. The number of manufacturing jobs in the capital continued to increase until 1961, and London is in the forefront of such innovations as radio, T.V. and computer electronics.

To take full advantage of mechanisation and to have room for expansion, a modern factory needs much open ground. Land for building is generally cheaper on the outskirts of a city than in its congested centre. This partly explains the phenomenal growth of London, which has been likened to the uncoiling of an octopus.

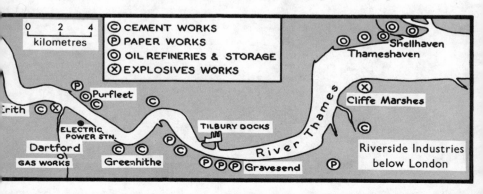

0   2   4
kilometres

Ⓒ CEMENT WORKS
Ⓟ PAPER WORKS
◎ OIL REFINERIES & STORAGE
⊗ EXPLOSIVES WORKS

Shellhaven
Thameshaven

Purfleet

Erith

ELECTRIC POWER STN.

Dartford
GAS WORKS

Greenhithe

TILBURY DOCKS

Cliffe Marshes

River Thames

Gravesend

Riverside Industries
below London

Spreading 'tentacles' of built-up land, with factories and housing estates all jumbled together, engulfed thousands of acres of fertile land all around the capital. (*10*) Make a labelled sketch of the photograph showing clearly where former fields have been 'invaded' by buildings.

For the most part the new estates are far pleasanter places in which to live than the grimy, congested central districts, but sprawling suburbs are a mixed blessing. As well as the rush-hour traffic problems, and the waste of up to three hours each day travelling to and from work, Londoners find it increasingly difficult to enjoy the pleasures of the open country.

Since 1935, attempts have been made to limit the growth of London by imposing a 'Green Belt' in which building is prohibited. Inducements (such as reduced rates) have also been offered to manufacturers to build elsewhere. In particular, 'New' and 'Expanded' Towns have been established on the outskirts of London to accommodate its 'overspill' population. These towns are not just outlying suburbs. They include workplaces and good shopping centres for their residents, thus relieving congestion in the central London area.

	City	Rest of the County of London	Total Greater London
1801	128 000	831 000	1 117 000
1841	124 000	1 826 000	2 239 000
1881	51 000	3 780 000	4 770 000
1921	14 000	4 471 000	7 488 000
1931	11 000	4 386 000	8 216 000
1951	5000	3 343 000	8 348 000
1961	5000	3 196 000	8 183 000
1971*	4000	2 734 000	7 614 000

* Figures are for the same areas, and *not* for the new (1965) GLC areas, which have slightly different populations.

It is now clear that efforts to decentralise London have been too successful. Between 1961 and 1975 the capital lost no fewer than 489 000 jobs, i.e. *34% of its manufacturing employment*. Hundreds of firms moved out, including many expanding and progressive enterprises. Scores of others were eliminated by redevelopment projects in older districts such as Stepney and Poplar. Parts of central London have therefore been drained of industry and skilled labour. One factory or warehouse in five lies empty, unemployment rates exceed 15% and there is an atmosphere of neglect and decay. Despite outward migration housing needs remain desperate and 350 000 dwellings are sub-standard.

To combat inner city decay the GLC (Greater London Council) made plans in 1978 to spend nearly £1000 million on new roads, houses, factories, services and general amenities. The aim is to reverse recent trends by drawing industrial jobs back to London.

The figures in the notes above show the changes in the population of different parts of London since 1801. (*11*) Use them to show by three separate graphs the population changes in (*i*) the City, (*ii*) the rest of the County of London and (*iii*) Greater London. (All of these districts are shown in this map. (*12*) What other counties are partly included in Greater London?) (*13*) Give reasons for the population changes shown in your graphs. (*14*) Do you think the trends of these graphs will continue in the same direction in future years? (*15*) If 'wedges' of land within the Green Belt were earmarked for industrial and residential use, what problems would be made (*a*) easier, and (*b*) worse?

The Greater London Area

THIS IS LONDON . . .

one of the world's largest cities, and the home of one in eight of Britain's population;

seat of the British Government, home of the Monarchy and centre of the British Commonwealth of Nations;

our most important port, and largest industrial city;

one of the world's chief banking and marketing centres;

home of Britain's largest university, art gallery, museum and library;

an historic city.

The mere fact that London *is* all of these things is sufficient to draw a multitude of people to her every year. As the headquarters of thousands of organisations, London attracts a constant stream of visitors on every conceivable sort of business. Scores of thousands flock to London, too, to see the pageantry of great occasions of State and great national sporting events like the Cup Final, Boat Race, two of the Test Matches, tennis championships and Rugby Football international matches. ((*16*) *Can you name all the places where these events take place?*)

Tourists and sightseers from all over the world come to London to visit famous places and buildings, some of which are sketched here: (*17*) how many can you recognise? Many visitors from overseas arrive in London by air, for air routes converge on the city from all over Europe and from every continent. There are four airports on the outskirts of London handling international traffic and an odd feature of air travel to London is that because of congested roads it often takes almost as long to reach central London from the airports as the whole of the time spent in flight.

Partly for this reason air travel within Britain has never developed on a scale comparable with that, for example, in the United States. (*18*) What other reasons are there? Would a large-scale helicopter service solve the problem?

The constant influx of visitors to London is an important source of wealth to a multitude of hotels, cafés, restaurants and shops. Many of the principal shopping streets are in the 'West End', e.g. Oxford Street, Bond Street, Regent Street and Piccadilly. Here, too, all jumbled together amongst attractive shops, are many theatres, cinemas and clubs, for this is the social district of London. Notice that it lies within handy reach of the main railway termini. (Map, page 182.)

There are districts similar to London's 'West End' in most very large modern cities. Other parts of such cities become chiefly concerned with commerce, or industry, or government, and the main residential districts are also often separate. (*19*) In London many of the great government departments, e.g. the Foreign Office, are in and around Westminster: what parts of the capital deal mainly with the other functions just mentioned? (*20*) What would you expect to see if you went to the following places: Threadneedle Street, Harley Street, Piccadilly Circus, Regent's Park, Whitehall, Drury Lane, King's Cross, Fleet Street, Tilbury, Smithfield, The Mall?

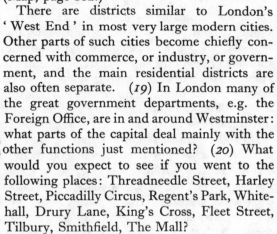

Not all of modern London is noise and bustle. Some of the ancient common lands have remained public open spaces by long usage. There are also famous Royal Parks such as Hyde Park, Greenwich Park and Kensington Gardens, and grounds which formerly belonged to country mansions. It would require three times more open space, however, to provide adequate recreation grounds for all Londoners, and other demands for land for houses, schools, hospitals, road improvements, etc., make it unlikely that such space will ever be found *within* London. Thus it is very desirable to preserve the 'green belt' of open country around Greater London referred to on page 186. (*21*) Are there similar land use problems in other parts of Britain?

189

CHAPTER 16

South-East England

THIS PART of Britain is a land of contrasts. A journey across it takes one past towering chalk cliffs, forested ridges, flat desolate marshes, rolling downs and rich fertile vales—all within a few miles. The reason for this remarkable variety of scenery may be seen in the map and geological diagram opposite.

Within the past 60 000 000 years the sedimentary rocks of south-east England were forced upwards to form an elongated dome, or anticline. The uppermost stratum was a thick deposit of chalk, which originally completely covered the dome. Through millions of years the top of the anticline has been eroded away by streams and the weather, so that the rocks lying below the chalk have been exposed. The chalk and sandstone strata now stand out as hilly ridges, and the clay sediments form vales. (*1*) Write down the reasons for this. (If you cannot, revise pages 28–29.)

(*2*) What different kinds of rocks, and how many ridges are crossed on a journey from Biggin Hill to Beachy Head?

The central sandstone ridge, known as Ashdown Forest, together with the horseshoe-shaped Wealden Clay Vale which encloses it, are called The Weald (see the block diagram).

(*3*) Make a large copy of this diagram, and after a careful comparison with the map complete the diagram by adding the following labels:—

NORTH DOWNS   RIVER THAMES
SOUTH DOWNS   VALE OF SUSSEX
COASTAL PLAIN   VALE OF HOLMESDALE
WEALDEN CLAY VALE   GREENSAND RIDGE[1]
FOREST RIDGES

Next, compare the diagram with that at the bottom of page 103. (*4*) What similarities and contrasts can you see between the geology and relief of south-east England and the Lake District? Can you account for the differences?

Now look at the course taken by the rivers in south-east England. Most of them rise in the forested sandstone ridges of the

[1] To the south of the Weald the Greensand Ridge is not so prominent, and is not shown on the map.

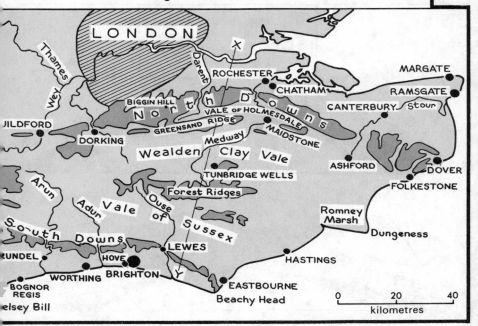

Weald, which in places reaches 230 metres. They flow either northwards to the Thames through gaps in the North Downs, or southwards through gaps in the South Downs to the English Channel.

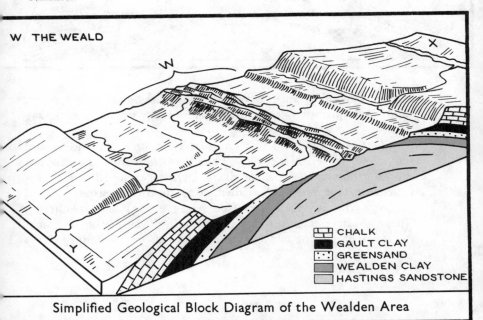

Simplified Geological Block Diagram of the Wealden Area

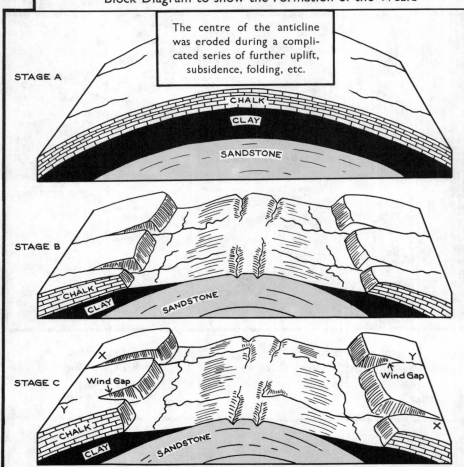

STAGE A

The centre of the anticline was eroded during a complicated series of further uplift, subsidence, folding, etc.

CHALK
CLAY
SANDSTONE

STAGE B

CHALK
CLAY
SANDSTONE

STAGE C

X
Wind Gap
Y
CHALK
CLAY
SANDSTONE
Y
Wind Gap
X

The block diagrams above show how these gaps were formed, and also explain why some of the rivers today follow such zig-zag courses.

Stage A. The anticline has just been uplifted, and is still completely covered by chalk. The rivers drain down the southern and northern slopes. Such rivers, having developed as a consequence of the uplift, are called **consequent streams**.

Stage B. River erosion (see pages 22–23) and weathering (page 18) have eaten through the chalk in the centre of the dome, to reveal sandstone and clay sediments below. Tributary rivers

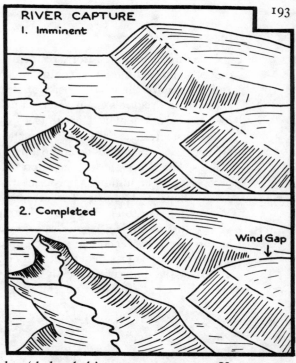

which have developed on the impervious clay join the main consequent streams more or less at right angles. These tributaries are called **subsequent streams.** Notice in this diagram, too, that the more resistant chalk is sticking out on the landscape to form escarpments, through which the consequent streams are cutting gaps.

Stage C. This stage represents conditions to-day. The main point here is that a tributary of consequent stream *X* has eaten so far back along the clay vale that it has ' beheaded ' consequent stream *Y*. This process, known as **river capture,** has played an important part in helping to fashion the present pattern of rivers in many parts of Britain.

Can you see, in the map below, where the River Darent has had its head-waters captured by the Medway? (5) Copy the map, show by a dotted line the probable former course of the Darent, and mark with a cross where you think river capture took place.

Bereft of its former headstreams, a 'beheaded' river carries a much reduced volume of water. If it shrivels so much that it no longer occupies the trench it had formerly cut through the escarpment, a **wind gap** is formed.

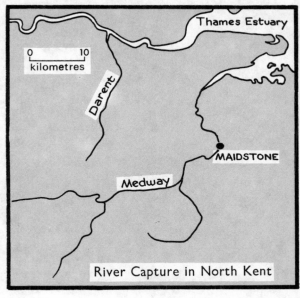

River Capture in North Kent

Relaxing at a coast resort, residents and visitors alike revel in the prolonged summer sunshine of south-east England. Kent and Sussex are the nearest parts of Britain to the Continent, and amongst the region's main assets are the hot, sunny summers of its 'semi-Continental' climate (see page 53). (6) What are the other characteristics of such a climate? Can you explain them? (If not, revise pages 50–53 and 62–63.) Huge crowds of holiday-makers, like those shown below on the beach at Ramsgate, bring prosperity every summer to proprietors of hotels, boarding houses, cafés, shops and amusement arcades. Similar crowds flock to Margate, Folkestone, Brighton, Hastings and Eastbourne.

The sort of crops a farmer can grow, and the kind of livestock he can best rear, depend very much upon the relief of his land and on the type of soil that covers it. As both these are largely deter-mined by the underlying rock it is easy to see why farming is very varied in Kent, Sussex and Surrey.

(7) Use the notes below to make a farming map of south-east England (similar to that for the lowlands flanking the Pennines on pages 136–137). A large outline copy of the map on page 191 will give you a start.

FARMING IN SOUTH-EAST ENGLAND

The Sandstone Ridges	The Clay Vales	The Chalk Escarpments
Soils on the Forest Ridges and the Greensand Hills are infertile. ((8) Why?) Saxon settlers found the Ridges covered with woodland and heath, left them uncultivated, and called the district the 'Weald'—meaning forest. Apart from DAIRYING in the valleys, much of the High Weald is still comparatively undeveloped as a farming area.	The Wealden and Gault Clay Vales are fertile —but damp. ((9) Why?) Grass grows well, and DAIRYING is important. At the fringe of the vales the clay has become mixed with chalk or sand to give very fertile loam soils on which large quantities of FRUIT (apples, cherries, soft fruit, etc.), HOPS and VEGETABLES are grown. Over one-third of Britain's fruit is grown in Kent and Sussex—mostly in Kent. The main hop-growing districts are around Farnham, Maidstone and Tonbridge, in the Vale of Holmesdale.	Soils are thin on the chalk ((10) Why?), but the covering of springy turf makes excellent pasture for SHEEP. RACEHORSES are bred and trained on the open country of the Downs, e.g. at Goodwood and Epsom. Both escarpments are smeared with 'clay with-flints' (see page 175), on which woodland and MIXED ARABLE farms are found. Much LAMB is produced for the London market. Parts of the North Downs lie on the southern fringe of Greater London, and here housing competes with farming for the use of the land.

Romney Marsh		
The shingle and alluvium have been drained to give rich, moist pastures. SHEEP are fattened here during the summer—the Marsh has the highest number per acre in Britain— but the flocks are moved in winter to the drier soils of the South Downs. Some of the big BULB growers of Lincolnshire have recently taken up land in Romney Marsh. The growing of GRASS FOR SEED is another specialised industry.		

Large quantities of dairy produce, fruit, hops and vegetables are sent to London and the seaside resorts. Every effort is made to produce as large a yield as possible. For such *intensive farming* much skill and labour is required. The soil needs constant attention—hoeing, weeding, manuring—and hops and fruit trees are sprayed against disease. Elaborate trellis supports are built to hold the climbing hop-plants. Large numbers of casual workers are employed for fruit-picking. Hops are now picked mostly by machine.

The coast of Sussex and Kent provides many fascinating examples of shore-line erosion and deposition. (Pages 38–40.) The results of both these processes can be seen in this photograph of St. Margaret's Bay, near Dover. The projecting chalk cliffs are gradually receding under the constant attack of the waves. As the chalk crumbles hard nodules of flint are washed out, and these are collecting to form a pebble beach in the Bay. (*11*) What was the state of the tide when this photograph was taken? How do you know? (*12*) With the help of the diagram on page 38, draw a fully labelled sketch to illustrate the photograph.

The maps show how much the coastline of south-east England has altered during the past two thousand years. In Roman times Reculver, Richborough, Dover and Lympne were all ports of entry. Dover, where there is a gap in the chalk cliffs and a valley leading inland, was the most important of these. Routes from the Roman ports converged at Canterbury and followed Watling Street to London. The Romans built no ports along

Map A: Coastline and Ports in Roman Times

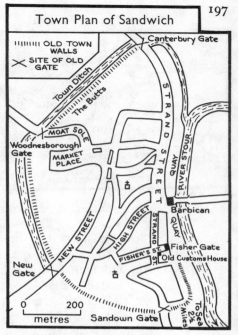

Town Plan of Sandwich

the coast west of Lympne : dense oakwood forests in the Weald hindered communications inland and the sheer chalk cliffs of the South Downs formed a barrier.

Reculver and Richborough were fortified ports guarding the Wantsum—a shallow channel used by Roman boats on their way to and from the Thames estuary. In time the Wantsum gradually silted up, and Saxon settlers, following the receding sea, established a port at Sandwich (*Sandwic*—the village on the sands).

In the Middle Ages Sandwich was one of the leading ports in Britain. It was one of the Cinque Ports—the others are shown on Map *B*—which held the monopoly of trade with the Continent. As the centuries went by, great shingle banks were built by longshore drift at Dungeness and Sandwich Bay. With the exception of Dover all the Cinque Ports gradually silted up and became quite impassable for ships of any size. To-day they are ' fossil towns ', left high and dry by the receding sea, but containing many reminders of their former existence as ports. (*13*) How many of these reminders can you see on the outline plan of Sandwich?

As Map *B* shows, several islands that existed in Roman times are now completely isolated from the sea. They include that district behind the North Foreland still known as the Isle of Thanet (*T* on map). A similar fate probably awaits the Isle of Sheppey (*S*). At Richborough, now $5\frac{1}{2}$ km inland, there are remains of only three walls which enclosed the Roman fort—the fourth side was protected by the sea.

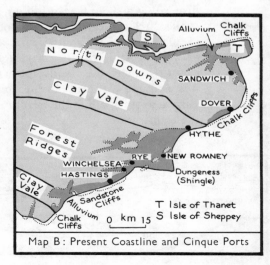

Map B : Present Coastline and Cinque Ports

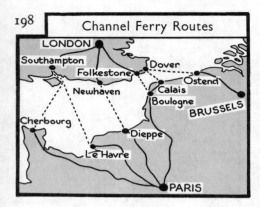

Channel Ferry Routes

Whilst other Cinque Ports fell into disuse, Dover retained its importance as the leading cross-Channel ferry-port. The traffic in passengers, goods and mail increased enormously with the construction of modern rail and road routes to London. Every year more than 5 million cross-Channel passengers stream through Dover's specially constructed quays, railway stations and Customs sheds. There are regular daily sailings. Similar facilities exist at Folkestone, Newhaven, Southampton and Harwich (see page 166) and a special train-ferry runs between Dover and Dunkirk. To help carry an ever-expanding holiday traffic, Britain's first hoverport has been built at Pegwell Bay. Huge SRN hovercraft, carrying 278 passengers and 36 cars, cross from Pegwell to Calais in 35 minutes.

In addition to its trade as a ferry port, Dover also exports coal from the small East Kent coalfield. An odd feature of this field is that it is entirely ' concealed ' (see page 112). Coal was first discovered here in 1890 during borings in connection with the Channel Tunnel project. An overhead cable railway was built to take coal from the collieries to Dover. Both the railway (now disused) and the pit-head buildings of the three collieries look quite out of place in what is otherwise an entirely rural setting. No ugly coal-tips mar the landscape, as waste rubble is spread out horizontally to avoid spoiling the beauty of the countryside.

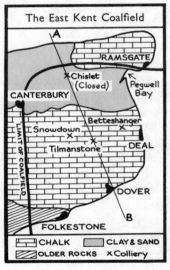

The East Kent Coalfield

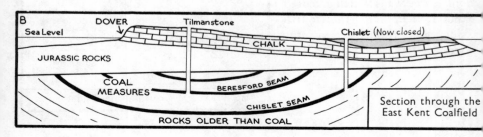

Section through the East Kent Coalfield

(*14*) The rock immediately below the coal here is Carboniferous Limestone: where is the nearest place at which outcrops of this rock are found?

Output from East Kent forms only one per cent of Britain's total coal production, and has given rise to no large-scale industries. In the Rochester–Gillingham–Chatham conurbation on the lower Medway, however, are a variety of industries including:—

Marine engineering	Vegetable bottling	Jam manufacture
Paper making	Oil refining	Brewing
Fruit bottling	Cement manufacture	Making farm machinery
Brick making	Naval shipbuilding	Steelmaking (small scale)

(*15*) Which of these industries do you think were influenced by (*a*) local agricultural production; (*b*) local port facilities; (*c*) local geology?

At Brighton and Ashford there are railway engineering works: both these towns owe the growth of these industries to their nodal position on the former Southern Railway network. Brighton has grown very rapidly during the past century (see page 262) and its miscellaneous industries include electrical engineering (e.g. making automobile parts, electric fires, switchgear, meters); radio engineering; plastics; clothing; confectionery; and brewing. Brighton has a splendid new marina and a new conference centre, and universities have been established at Brighton, Guildford and Canterbury.

Chatham was originally a naval port, built to guard the Thames estuary and the Straits of Dover. As the chief gateway to England, the south-eastern counties are strewn with the remains of many land fortifications. Romans, Saxons, Normans, Tudors— they all built castles. Many of them, like the Norman fort at Lewes in the sketch, were located in strategic gaps through the chalk escarpments on routes leading to London. Gap towns which grew up under protection of the forts were natural route centres, where in time markets developed for the surrounding countryside. In some cases gap towns grew up in the fertile clay vales at the entrance to the gaps, e.g. Canterbury, Ashford, Maidstone. (*16*) From the map on page 191 how many gap towns can you find in the North and South Downs?

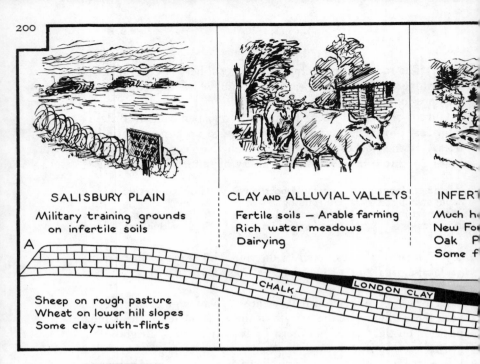

SALISBURY PLAIN

Military training grounds
on infertile soils

Sheep on rough pasture
Wheat on lower hill slopes
Some clay-with-flints

CLAY AND ALLUVIAL VALLEYS

Fertile soils — Arable farming
Rich water meadows
Dairying

INFERT

Much h
New Fo
Oak P
Some f

A

CHALK   LONDON CLAY

CHAPTER 17

The Hampshire Basin

THESE MAPS and diagrams show many of the main facts about
the geography of the Hampshire Basin. (1) Study them carefully and then write a geographical account of a journey from
Devizes to Freshwater Bay. In your description pay particular
attention to the geology, scenery, agriculture and other activities
in the districts through which you would pass.

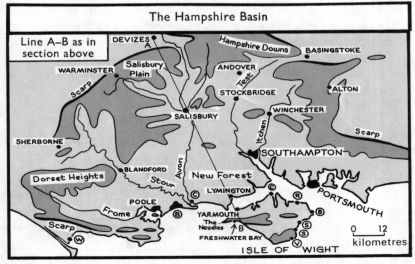

Map A

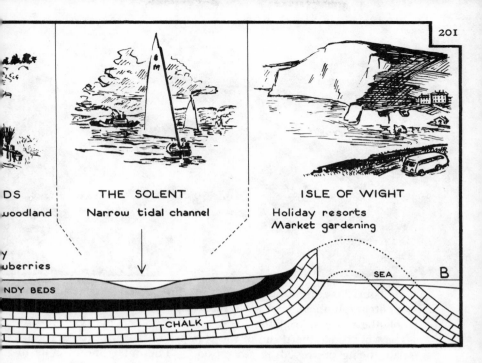

THE SOLENT
Narrow tidal channel

ISLE OF WIGHT
Holiday resorts
Market gardening

DS
woodland

y
uberries

NDY BEDS

CHALK

SEA

B

It is thought that the Isle of Wight was once joined to the mainland. (*2*) What evidence for this can you deduce from the maps? (*3*) What appears to have been the drainage pattern at that time?

The Hampshire Basin contains fine stretches of heath and woodland. In 1978 oil was found at Wytch Farm in substantial quantities—equal, for example, to that of the Auk field under the North Sea. Its exploitation has been opposed by conservationists on the ground that rare flora and fauna on heathland at Wytch would be disturbed. (*4*) What arguments are there for and against oil production in these circumstances?

On the coast are many thriving resorts, favoured by sheltered positions ((*5*) *Why?*) and a mild climate. (*6*) How many of those indicated on Map *A* can you name?

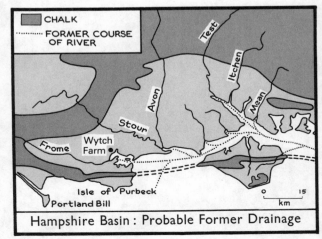

CHALK
FORMER COURSE OF RIVER

Test
Itchen
Meon
Avon
Stour
Frome
Wytch Farm
Isle of Purbeck
Portland Bill

0   15
km

Map B

Hampshire Basin : Probable Former Drainage

Tourists are attracted, too, by fine coastal scenery. The photograph above shows the remarkable coast at Lulworth Cove on the south side of the 'Isle' of Purbeck. The 'Isle' forms part of the southern rim of the Hampshire Basin, which is gradually being destroyed by sea erosion. The outer line of cliffs in this picture consist of relatively resistant Purbeck Limestone. Where this has been breached, the sea has quickly eroded deep holes in the softer clay beds behind.

The first stage in this process has occurred at Stair Hole, where

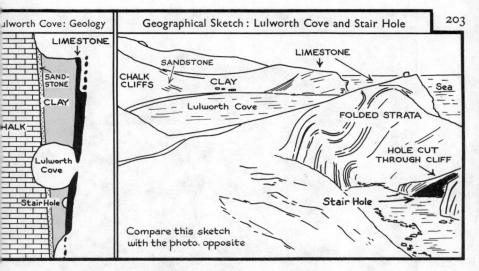

Lulworth Cove: Geology

LIMESTONE

SAND-
STONE

CLAY

CHALK

Lulworth
Cove

Stair Hole

LIMESTONE

SANDSTONE

CHALK
CLIFFS   CLAY

Lulworth Cove

Sea

FOLDED STRATA

HOLE CUT
THROUGH CLIFF

Stair Hole

Compare this sketch
with the photo. opposite

waves have cut two small holes through the base of the limestone
cliffs.   At Lulworth Cove, where the limestone barrier has been
completely broken through, the sea has scooped out a deep bay
in the clay, and begun to attack the more resistant sandstone and
chalk strata farther inland.

Farther east, the chalk escarpment has been completely demolished by sea erosion for a distance of over 25 km (see maps
on pages 200–201).   The eastern end of this gap is marked
by The Needles, at the western extremity of the Isle of Wight.
The lower photograph shows the line of isolated rock pinnacles,
called **stacks,** which have been formed here by the scouring
action of the waves (page 39).   (7) Make a large copy of the
sketch below, adding the following labels in their appropriate
places :

STACK	CAVE	JOINT
UNDERCUTTING	ARCH	ROOF COLLAPSED

(8) By stringing these words together in the right order you can
describe fully the formation of a stack.

Now look again at the photographs.   (9) In which direction was
the camera pointing when each was taken?   (10) What similarities
can you see between the dip of the strata at Stair Hole and at The
Needles?   (11) What characteristics of chalk downland
can you see in the lower photograph?   (12) What
evidence can you find in it suggesting
that wave erosion is still taking place
at The Needles?

Southampton is the most important commercial port on the south coast and a busy cruise and ferry terminal. The Old Docks, and part of the city, are shown in this photograph. *(13)* How many of the docks can you pick out? Where is the railway terminus?

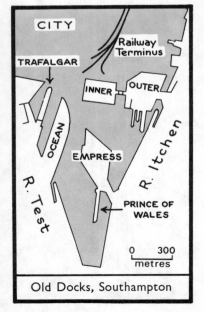

Old Docks, Southampton

Because there is only a comparatively small rise and fall of the tide in Southampton Water, there is no need here for lock-gates across the entrances to the docks. (Compare these maps and photograph with those of the London docks on pages 180–181.) Southampton tides are peculiar in that there are four high tides a day instead of the normal two. This is one of the port's main natural advantages, for ships of all sizes can enter or leave the docks for long periods every day.

Other factors which help to explain Southampton's growth into a leading port

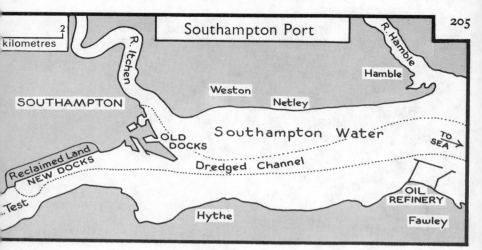

Southampton Port

include:—

its double waterfront, overlooking both the Rivers Test and Itchen; the soft clays and gravels which made dock excavation relatively simple; its position on the south coast which made it a convenient port of call for Dutch, French and German trans-Atlantic liners; its excellent road and rail links to London (81 minutes away by fast trains) and with the rest of England (see p. 201). (*14*) Why is it sometimes called ' The Outport of London '?

Southampton's varied port facilities include large dry-docks, a magnificent ' Ocean Terminal ' for cruise-liner passengers, a fast-growing vehicle ferry service to Cherbourg and new container docks. The once-famous trans-Atlantic passenger trade has faded ((*15*) *Why?*) but since 1969 container and ' ro–ro ' traffic through Southampton has jumped from 0.08 to 2.9m. tonnes p.a. South Hampshire (i.e. mainly Southampton–Portsmouth) has been officially designated a growth centre of the first priority, mainly to take the strain off the Greater London region. (*16*) Explain why Southampton has many advantages for industrial expansion now that Britain has entered the Common Market.

SOUTHAMPTON'S CARGO TRAFFIC

Some Important Imports			Exports
ruit	Flowers	Wool	A great variety of manufactured goods, especially vehicles and engineering products from the Midlands and W. London. Road links to the Midlands have recently been improved.
Meat	Hides	Jams	
Vegetables	Skins	Bananas	
Grain	Timber	Petroleum	
pecially equipped docks handle the many perishable imports.			

The huge oil refinery at Fawley includes fast-growing petro-chemical works. Pipe-ines supply ethylene to the I.C.I. works on Severnside, aviation spirit to a depot near London Airport and fuel to several gas and electricity power stations. The 24m. tonnes of petroleum landed at Fawley each year constitutes 86% of Southampton's total goods raffic.

Salisbury Plain

RIDGE WAY

Stonehenge

PORT WAY

R. Test

R. Itchen

PILGRIM'S WA

SALISBURY

R. Avon

WINCHESTER

0 6 12
kilometres

ANCIENT TRACKS

Inland from Southampton lie the ancient cathedral cities of Salisbury and Winchester. Both are gap towns, with similar origins and growth, for both lie on valley routes leading across the chalk rim of the Hampshire Basin.

Salisbury is a natural route focus, where five tributary streams of the River Avon converge. There is plenty of evidence that men lived in the Salisbury district at a very early date. On the outskirts of the town is Old Sarum, a prehistoric hill-top fortress, and other camps, earthworks and burial chambers are scattered over the adjoining plain. The burial grounds, probably 4000 years old, are marked by mounds called **tumuli,** or **barrows.** Many of these have been excavated and their contents—skeletons, weapons, tools—studied by archaeologists in an effort to build up a clearer picture of how our remote ancestors lived and worked. By plotting the position of these finds on maps we can also get a good idea of *where* they lived. Such maps suggest that prehistoric men in Britain made their homes on the chalk and limestone escarpments. These were less densely forested than the clay vales ((*17*) *Why?*) and easier to clear with primitive implements. Hill sites also afforded protection, and travel was easier on the ridges than over the marshy lowlands.

EACH DOT MARKS A FIND ABOUT THE SAME AGE AS STONEHENGE

S. and S.E. England: Neolithic Finds

Along the ancient hill-top trackways which converge on the Salisbury Plain, tribesmen probably travelled to religious ceremonies at the mysterious stone circles at Avebury and Stonehenge.

Both Salisbury and Winchester were important towns during the Roman occupation of Britain, and during the Saxon invasions which followed. Winchester's chief importance lay in its control of the main route from Southampton to London. ((*18*) *What rivers does this route follow?*) Six Roman roads met there, and the Saxons later made it the capital of their Wessex kingdom. Today Salisbury and Winchester are the county towns of Wiltshire and Hampshire respectively, and important marketing centres for the surrounding countryside. At Winchester is a famous school, founded in 1382 by William of Wykeham.

The strongly fortified naval base at Portsmouth is a reminder that southern England has often been liable to attack from the Continent by sea. The large drowned harbour here, sheltered from westerly gales by the Isle of Wight, has been used by the Navy since the early Middle Ages. (*19*) From the map below can you see what advantages for defence the harbour site afforded?

The sketch above shows Nelson's flagship *Victory*, which is preserved at Portsmouth in dry-dock.

With over 200 000 inhabitants, Portsmouth is the largest town in the Hampshire Basin. There are marine and general engineering works. At the naval depot ships are refitted and supplied with food, clothing and ammunition for lengthy voyages. Sailors are trained at shore establishments, and many of them have homes in the town.

Portsmouth has a major office complex containing the Post Office and the Inland Revenue computer centres. In nearby Havant a growing satellite development of factories and houses provides 24 000 jobs, mostly in engineering and varied light craft industries.

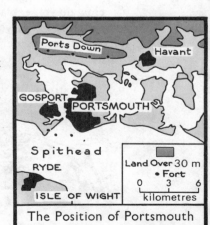

The Position of Portsmouth

CHAPTER 18

The West Country

(A) THE SOUTH-WEST PENINSULA

A **peninsula** is a piece of land which has water on three sides, and is thus particularly subject to coastal erosion; yet the South-west Peninsula forms one of the oldest parts of the British Isles. The maps show why the land here has been able to withstand the relentless attack of the waves for perhaps 350 million years (see also page 9).

The maps also reveal other important facts about the geography of this part of Britain. Study them carefully, and see how many of the following questions you can answer correctly:—

(*1*) What relation can you see between the relief and the path followed by the railways ?

(*2*) Which moorland area differs geologically from the others ?

(*3*) The cliffs of the Lizard Peninsula are made of a metamorphic rock called serpentine: which cliffs do you think they more closely resemble in hardness and appearance, those at the Land's End or at Ilfracombe ? (Give reasons.) (See page 11.)

(*4*) Why are most of the principal towns on the coast ?

(*5*) What natural advantage favoured the growth of each of the following ports: Dartmouth, Plymouth, Fowey, Falmouth, Padstow, Bideford, Barnstaple ?

(*6*) Study the geological map and suggest, with reasons, what rock you would expect to find in the Isles of Scilly.

(*7*) How do you account for the distribution of rainfall ? (See also pages 56 and 62–63.)

(*8*) How do you explain the pattern of the isotherms in January and July ? (If you find this difficult, revise pages 52–53.)

(*9*) Which river marks, for much of its length, the county boundary between Devon and Cornwall ? (Use your atlas.)

(*10*) Many trees in west Cornwall are shaped like the one sketched below—*why* ?

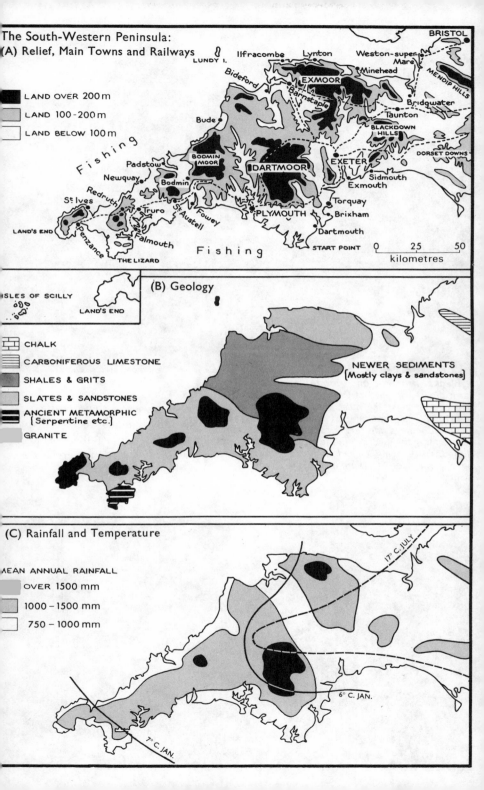

The South-Western Peninsula:
(A) Relief, Main Towns and Railways

LAND OVER 200 m
LAND 100-200 m
LAND BELOW 100 m

BRISTOL
Ilfracombe
Lynton
Weston-super-Mare
Minehead
LUNDY I.
Bideford
EXMOOR
Barnstaple
MENDIP HILLS
Bridgwater
Bude
Taunton
BLACKDOWN HILLS
DORSET DOWNS
Fishing
Padstow
BODMIN MOOR
DARTMOOR
EXETER
Newquay
Bodmin
Sidmouth
Redruth
Exmouth
St Ives
Truro
St Austell
Fowey
Torquay
LAND'S END
Penzance
PLYMOUTH
Brixham
Falmouth
Dartmouth
THE LIZARD
Fishing
START POINT

0   25   50
kilometres

ISLES OF SCILLY
LAND'S END
(B) Geology

CHALK
CARBONIFEROUS LIMESTONE
SHALES & GRITS
SLATES & SANDSTONES
ANCIENT METAMORPHIC [Serpentine etc.]
GRANITE

NEWER SEDIMENTS [Mostly clays & sandstones]

(C) Rainfall and Temperature

MEAN ANNUAL RAINFALL
OVER 1500 mm
1000-1500 mm
750-1000 mm

17° C. JULY
6° C. JAN.
7° C. JAN.

Bleak, heather-clad granite moorlands form the 'backbone' of Devon and Cornwall, and are amongst the most desolate districts in Britain. In great contrast are the pleasant sheltered valleys and fertile sandstone plains surrounding the moors. In the picture below, daffodils are being picked in late February on a farm near St. Dominic, in the Tamar Valley.

On page 9 we saw that granite is a crystalline igneous rock. The crystals—quartz, mica and felspar —are very resistant to erosion,

but in time are broken up by the chemical weathering of the weak acids in rainwater. Some of the chemicals are washed away in solution, but quartz grains and clay remain behind as an insoluble residue. The bogs on Dartmoor—a danger to unwary hikers— were formed in this way. Granite moorlands have a rounded appearance. Their wide, open valleys are often marshy. ((*11*) *Why?*) Soils are sandy, thin and infertile, and in higher places bare rock ' tors ' show through—as at Manaton Tor in the photograph opposite. Some sheep are reared, and a few wild cattle wander over the unfenced hills, but cultivation on the latter is impossible. (*12*) What else, besides poor rocky soils, prevents it?

In spite of the large tracts of barren land, farming is the main means of livelihood in Devon and Cornwall. The climate is particularly equable on the lower coast plains, and although winter snowfall is heavy on the moors, especially Dartmoor and Exmoor, it is sufficiently rare in coast towns like Penzance and Falmouth to cause considerable excitement. The mild conditions permit palms, draecenas, and other sub-tropical plants to survive outdoors, and even bananas will grow (but not ripen).

The heavy rainfall and equable climate foster rich pastures on the fertile sandstones, and dairying is very important. Butter and cream are produced in milk factories at such places as Sancreed (near Penzance) and Lostwithiel. Except for Ireland, no other part of Europe is so well suited to dairy farming as Devon and Cornwall. Farming is mixed, ' dredge corn ' (oats and barley sown together) being grown for cattle food, and skimmed milk is fed to pigs. Grain growing, however, is a chancy business because the wet, mild southwesterly airstream encourages damaging rusts and mildews. Beef cattle are also reared, especially in north Devon, where the ' Devon ' breed originated. In west Cornwall and the Isles of Scilly, where winters are exceptionally mild and frosts rare, large quantities of early spring vegetables and flowers are grown. Loads of broccoli, potatoes, daffodils, narcissi and anemones are sent by road and rail to the London markets. Fruit growing is also important, especially apples, pears and plums. Devonshire cider orchards are famous.

The ria coastline of Devon and Cornwall contains many excellent sheltered harbours. Most of the inlets once had flourishing fishing fleets. Today the inhabitants mostly cater for tourists. Mousehole (above) is a favourite haunt for sightseers. The drowned estuary of the River Fal (below) forms one of Britain's finest natural harbours. The photograph shows a small oil-tanker in dry-dock at Falmouth—a ship-repairing port which has specialised in work of this type of vessel. Now the dockyard is threatened with closure because it is too small to be economically viable. Falmouth's harbour could handle the largest passenger and cargo vessels with ease, yet it has never developed as a passenger or commercial port. (13) How has Falmouth's geographical position helped to keep its beauty unspoiled by development?

The indented coastline helps to explain why the people of the south-west have an age-old tradition of sea-faring and fishing. Most of the ships which destroyed the Spanish Armada sailed from West-country ports and were manned by West-country crews. Now, centuries later, many Devonshire and Cornish youths still join the Royal Navy and the Merchant Marine. There is an important naval base and dockyard at Devonport, and a Royal Naval College at Dartmouth.

Soon after Cabot had charted Newfoundland and the St. Lawrence estuary, Devonshire fishermen were crossing the Atlantic in tiny craft to visit the rich fishing grounds there. In the Middle Ages much dried and salted fish was exported to Catholic countries like Italy: hence the old Cornish rhyme:—

> " *Here's a Health to the Pope! may he live to repent*
> *And add just six months to the term of his Lent,*
> *And tell all his vassals from Rome to the Poles*
> *There's nothing like pilchards for saving their Souls!* "

In recent decades the fishing industry in the south-west has steadily declined: the main reasons include:—

1. Fewer fish—especially pilchards—are breeding or migrating off the South-west Peninsula.
2. Much larger and better-organised fishing fleets have grown up, based on the North Sea ports.
3. Devon and Cornwall are far from markets in the big centres of population in industrial Britain. (Compare the distance of Grimsby and Newlyn from London and Birmingham.)

Nevertheless, such places as St. Ives, Newlyn, Mevagissey, Looe, Plymouth and Brixham still have small fishing fleets. Falmouth is also visited by East European fish-factory ships, which buy mackerel from British trawlers. West Country boats also land considerable quantities of shellfish (oysters, crabs and lobsters), for sale to hotels and restaurants, or for processing at quayside factories.

The fine ria harbours are ideal for boating and sailing, and there are pleasure boats, small boat-building firms, dinghy-yards and marinas on most of the larger creeks.

The equable climate, spectacular cliff and moorland scenery, and quaint fishing coves attract an annual influx of tourists to Devon and Cornwall. The money they spend provides many local inhabitants with their largest single source of income. Guest houses and large luxury hotels are often perched on the cliff tops at resorts like Torquay, Falmouth, St. Ives and Newquay. Most visitors arrive during the summer, but some—especially invalids—are attracted to the South-west in winter by the mild climate.

In places granite has decomposed, as vapours and gases from deep in the earth's crust have risen through the rock. The decayed granite which contains much china-clay (kaolin) and sand (quartz) is soft, white and crumbly, and is taken from huge pits like that in the photograph. The china-clay is washed out by high-pressure hoses and allowed to settle in tanks, whilst the sand is dumped to form gleaming white tip-heaps. Most of the china-clay from the St. Austell pits is taken to the ‘Potteries’ by lorry (and not, as formerly, via the Trent–Mersey Canal). Much is also exported through Fowey to the U.S.A. and Germany, for paper-making (see table below).

In places where granite has been unaffected by chemical decay, its great strength makes it an excellent building and monumental stone. Many Cornish houses are built of it (photograph, page 261) and Cornish granite from the Falmouth district has been used for many large constructions in Britain, including Waterloo Bridge. Various hard, tough rocks are quarried in a number of places in Cornwall for use as road stone and there is also an enormous slate quarry at Delabole. (*See photo.*) ((*14*) *Where else have we found granite and slate close to one another?* (*15*) *What is the explanation?*)

Round the granite masses of central and west Cornwall there are veins of mineral ores. Copper and tin have been produced there for thousands of years. The peak of Cornish mineral production was reached in the late 19th century. Thereafter, with the increasing depth of the mines, and growing competition from production in Australia, Nigeria, Bolivia and the easily worked alluvial tin deposits in Malaya, the Cornish mining industry virtually ceased. Following sharp

---

THE CHIEF USES OF CHINA-CLAY

Manufacture of:
Paper (as a ‘ filling ’ to give the paper its glossy surface);
High-grade pottery; Paint; Oil Cloth, Linoleum, Rubber; Medicinal Kaolin (e.g., poultices).
Also used as a base for toilet preparations and cosmetics: e.g. face powders, toothpastes; skin creams, etc.

Large white mounds like these dot the landscape in the St. Austell district of Cornwall. From a distance they have the appearance of snow-covered hills, but in fact are waste-tips from china-clay quarries.

increases in the price of tin in the 1960s, two new mines were opened near Truro, but their economic future remains precarious.

At Camborne there is a long-established School of Mines for training metalliferous mining engineers, and an engineering firm which specialises in mine-drilling equipment.

Cornwall was one of the zones in west Britain where the Celts settled in prehistoric times (see also page 223). Little affected by invading tribes in south-east Britain, the Cornish have much in common with their fellow Celts in Wales and Brittany: they spoke a language similar to Welsh, and almost identical with Breton, until late medieval times. Many Cornishmen have Celtic names, and Celtic place-names are very common, especially Tre- (place), Pol- (pool), or Pen- (headland).

The great quarry at Delabole, East Cornwall, where slate has been worked continuously since prehistoric times. Modern architectural interest in the beauty and durability of natural stone ensures a steady market for slate slabs, sawn building stone, crazy paving and hand-split roofing slates.

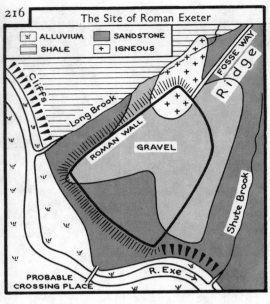

The Site of Roman Exeter

ALLUVIUM SANDSTONE SHALE + IGNEOUS

Cliffs
Long Brook
FOSSE WAY
Ridge
ROMAN WALL
GRAVEL
Shute Brook
R. Exe
PROBABLE CROSSING PLACE

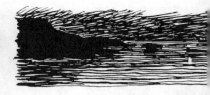

Like the rest of Celtic Britain, the far south-west was little influenced by the Roman occupation, the most westerly Roman town being EXETER (97 000). (*16*) Explain from this map why:— 1. it was easy to defend; 2. it was a convenient place to cross the River Exe; 3. it could be approached easily from the north-east; 4. it gave protection from flooding and provided dry ground on which to build.

In the Middle Ages Exeter became prosperous as a cathedral city, and as a market centre for the West-country woollen trade (see page 220). As this trade gradually declined, Exeter remained prosperous as a marketing place for the rich agricultural lands surrounding it. The map below shows why Exeter became a route focus, first for the early turnpike roads, and later for railways and modern roads. Now Exeter is the terminus of the M5 Motorway (*map p. 281*), and South Devon resorts are within weekend commuting distance of London and Birmingham.

The M5 is carried by a fine new by-pass to the south of the city, and this has eased congestion of holiday traffic. The new motorway link also favours the city's industrial estates, with their thriving light-engineering firms. Easier access will also benefit Exeter's expanding university and reinforce the city's role as the chief market and shopping centre in east Devon. No less than 38% of Exeter's workforce is in administrative and professional services.

TO BARNSTAPLE
R. Exe
TO BRISTOL & LONDON
TO OKEHAMPTON
TO SALISBURY & LONDON
Dartmoor
EXETER
Torquay
LAND OVER 200m
MAIN RAILWAYS
TO PLYMOUTH

The Location of Exeter

Exeter is an important tourist centre: (16a) what coastal resorts and moorlands are easily reached from the city?

PLYMOUTH (250 000) is the largest city in the south-west. The chief single source of employment is the naval dockyard at Devonport, which guards the western approaches to the English Channel. The largest warships can enter this fine, sheltered, natural harbour. The dockyard, which specialises in refitting missile cruisers, has been expanded and roofed in to allow work to continue in all weathers. Like Falmouth, Plymouth has never become a great port ((*17*) *Why?*) but its docks handle much fruit and vegetables imported from N.W. France and the Mediterranean. A 'ro–ro' ferry service to Roscoff (Brittany), opened in 1973, has benefited from Britain's joining the Common Market. Plymouth is also a port of call for cruise liners and is a fishing, marketing and major shopping centre. Recently established light industries (e.g. scientific instruments, potato crisps, infant foods, chewing gum, glass-fibre boats) employ some 10 000 people.

" It was evening when I went up to the Hoe. Darkness was falling. . . . Far below, the grey waters of the Sound lay smooth between sheltering cliffs. The breakwater was a faint line on the sea. There were lights on Drake's Island; lights on Mount Edgecombe; lights on a grey destroyer steaming towards the dockyard at Devonport. From the right came the distant sound of hammering against metal: a vigorous touch in this peaceful scene. . . . Beyond the breakwater a liner passed . . . her portholes like a string of pearls on the water. A ship's siren called at the mouth of the harbour. . . . Straight out of the sea, fifteen miles away over grey waters, snapped a light, the most famous light on the British coast: the light at the Eddystone.
" I am sorry for the man who can stand for the first time on Plymouth Hoe without a tingling of the blood." [1]

On this map find the spot where the viewer who wrote the above description was standing. (*18*) *Then copy the sketch above, adding the following labels:*—

DRAKE'S ISLAND   THE SOUND
BREAKWATER   PENLEE POINT
    EDDYSTONE
    MOUNT EDGECOMBE

(*19*) *Suggest an explanation for " the distant sound of hammering against metal " in the description.* (*20*) *What is the approximate compass course of the destroyer?*

[1] H. V. Morton,
In Search of England, op. cit.

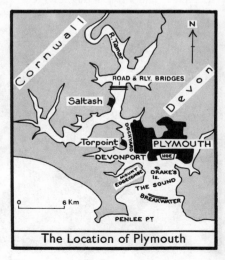

The Location of Plymouth

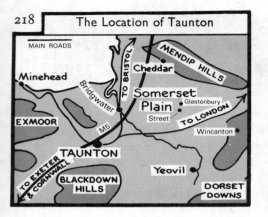

The Location of Taunton

MAIN ROADS

Minehead
TO BRISTOL
Cheddar
MENDIP HILLS
Bridgwater
Somerset
Plain
Glastonbury
EXMOOR
M5
Street
TO LONDON
Wincanton
TAUNTON
TO EXETER & CORNWALL
Yeovil
BLACKDOWN HILLS
DORSET DOWNS

The eastern boundary of the South-west Peninsula may be taken as the Mendip Hills and the western rim of the Hampshire Basin. The district shown in this map differs considerably from the region farther west. The chief contrasts are found in the flat alluvial land of the Somerset Plain and the limestone hills of the Mendips.

In many respects the Somerset Plain resembles the fenlands of East Anglia. Drainage on the impervious alluvium has always presented difficulties, and until recently large portions of the Plain were marshy and difficult to cross or cultivate. The main railway from Bristol to Penzance runs on an embankment for part of its way across the Plain, for flooding is still frequent in winter.

Lush grasses grow on the drained alluvium, for the Plain has a moist, equable climate similar to that of the coast lowlands of Devon and Cornwall. Dairying and fruit-growing are important: the Cheddar region has long been famous for its cheese, whilst Somerset's orchards and cider are equally well known.

Taunton and Bridgwater are the chief marketing centres on the Somerset Plain. (*21*) Why did Bridgwater grow up at that particular point on the River Parret? (*22*) Why is Taunton called the 'gateway' to the south-west?

Thick deposits of peat (decayed marsh vegetation) are found in parts of the Somerset Plain and are dug for fuel. (*23*)

Why have the peat blocks shown below been stacked so carefully? (*Hint: captio p. 236.*)

The enormous chasm of Cheddar Gorge in the Mendips is the most spectacular 'karst' feature of these limestone hills. At the foot of the Gorge is a labyrinth of underground caverns and pot-holes, with stalactites, stalag-mites and underground streams. (See also pp. 26–27.) The Mendips receive a substantial rainfall (map, p. 209): (24) Why, therefore, is the natural vegetation in this photo so scanty? On such rough herbage sheep-rearing is the only important type of farming.

Many small towns in the South-West have recently expanded light industries, mainly engineering and food processing. Examples include Yeovil (Westland helicopters), Wincanton (Horlicks), Appledore (shipbuilding), Bridgwater (engineering), Street and Glastonbury (leather and plastics). With the coming of the M5 such towns could change out of all recognition, but industrial growth in the South-West is hindered by a shortage of labour, especially as very few surplus farm workers are left on the land.

Avon Gorge and Clifton Suspension Bridge

(B) BRISTOL AND THE SEVERN ESTUARY

East of the Mendips, towards the western edge of Salisbury Plain, there are towns such as Frome, Westbury, Trowbridge and Bradford-on-Avon, which specialise in producing high-quality woollen fabrics. This district formed part of the once-famous West Country woollen manufacturing region, which extended north-eastwards along the line of the Cotswold Hills into Gloucestershire and Oxfordshire (see also page 172). Local wool supplies, water power and deposits of fuller's earth (used for washing the greasy raw wool) favoured the growth of a textile industry. Since the Industrial Revolution and the large-scale development of the West Yorkshire woollen manufacturing region, with its large coal resources, the West Country wool trade has suffered decline, except in the making of specialities. (25) Where else in Britain has this occurred?

Eight miles from the sea the River Avon follows a deep gorge through a small limestone ridge. Just above the gorge is Bristol, an industrial and university city and at one time the chief port on the west coast of Britain. During the Middle Ages much wool was exported through Bristol to south-west France, wines being imported from the Bordeaux region in return. Later the port was concerned with the American trade, was the main base of the British slave traders (*p. 139*) and handled much tropical colonial produce. (26) Study the list of Bristol's industries and suggest which may have originated at this period.

At one time Bristol was the main exporting centre for the English low-lands, but as ships grew larger this function passed to Liverpool. Modern

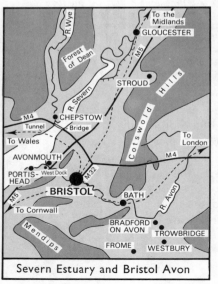

Severn Estuary and Bristol Avon

Aero-engineering	Aircraft construction	Boots and shoes
Engineering	Jam and pickle making	Printing
Brewing	Packaging	Wine bottling
Manufacture of cocoa and chocolate (Fry's)		Rope making
Cigarette and tobacco manufacture (Wills')		Brushes (Kleen-e-ze)

docks have been built downstream at Avonmouth, which is the **out-port** for Bristol. Avonmouth is specially equipped to handle frozen meat and dairy produce, grain, bananas and petroleum, and has chemical and metallurgical industries. The huge new West Dock (Royal Portbury) imports timber, and is a 'ro-ro' and container terminal. Port operations on the Avon are handicapped by the enormous rise and fall of the tide in the Severn estuary. *

Bristol has probably reached its zenith as a port (*see notes above*) but seems destined for further large-scale growth as an industrial centre. For many years the city's economic mainstay has been its aerospace and aircraft engineering industry—9000 workers are employed in the B.A.C. factory at Filton, famous for its Concorde project. Motorway links with London, the Midlands and Exeter (*map, p. 281*) make Bristol an ideal distribution and service centre, a fact emphasised by the huge office blocks which now dominate the city centre. (27) Use the following figures to draw bar diagrams of the main types of employment in Bristol:— office jobs (60 000); finance and professions (40 000); distributive trades (30 000); aero-space and aircraft (24 000); transport and communications (20 000); printing, paper and publishing (16 000); mechanical and electrical engineering (14 000); others (100 000).

Gloucester grew up as a market town and route centre when it was the lowest bridging point of the Severn. Now east–west traffic can avoid the city—(28) *how?* Gloucester is best-known today for its great aircraft industry.

* This big tidal range—up to 20 m at
?pstow—is probably caused by the funnel
?e of the estuary. As the incoming tide
?ls up-stream the water becomes in-
?singly confined, and piles up. At the
?hest or 'spring' tides a wall of water
?eral feet high rushes up the river. The
?e is called a **bore**, and is a serious
?drance to navigation. Its occurrence helps
?xplain why Gloucester, at first sight so
?urably placed in relation to the Midland
?n, never became a great port for that part
? England. The approach up-river to
?ucester is also shallow, and the ship-canal
?n Sharpness can take only small vessels.

CHAPTER 19

Ireland

THERE IS plenty of evidence to suggest that in earlier geological times Ireland was joined to the rest of the British Isles. The main connecting links are indicated below. (*1*) Study this map, compare it with the relief map on page 226, and then sort the following into correct geological pairs:—

Mountains of Donegal, Mayo and Connemara; Welsh Plateau; Scottish Highlands; Wicklow Mountains; Mourne Mountains; Hills of Devon and Cornwall; Southern Uplands; Mountains of Munster.

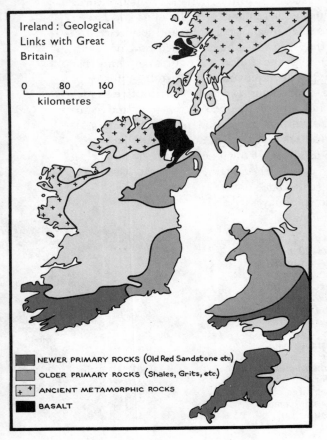

Ireland : Geological
Links with Great
Britain

0   80   160
kilometres

NEWER PRIMARY ROCKS (Old Red Sandstone etc.)
OLDER PRIMARY ROCKS (Shales, Grits, etc.)
ANCIENT METAMORPHIC ROCKS
BASALT

Additional evidence of a former link is that volcanic rocks similar to those which form the Antrim Plateau are found off the Scottish coast in the islands of Mull, Iona and Staffa. (*2*) Does the geology of the Isle of Man provide yet another clue?

The mountains of Ireland form a broken rim of highland around the low-lying Central Lowland. This plain is made of

Carboniferous Limestone, but is very badly drained because it is overlain in most places by deposits of impervious boulder clay. (See map, pages 64–65.) Rainwater has dissolved the underlying limestone in places to form hollows in the ground surface, which are filled by shallow lakes and peat-bogs. The Bog of Allen is 32 km across: notice in the maps on pages 226 and 237 how the railways skirt its edges. (3) What other links can you see between the railway network and the relief? (4) Where are the main rail centres in Ireland? *

Many roads across the Central Lowland run along the top of sandy ridges called **eskers**, which are believed to have been formed from sediment deposited by streams running beneath the ice-sheet which formerly covered most of Ireland. The higher hills show evidence of glacial erosion, for most of their thin soil-cover was scraped off by the ice and dumped in the valleys and on the Central Lowland. In the north some U-shaped valleys were flooded by the sea to form narrow inlets called loughs. (5) The similar inlets in the south are rias—how do these differ from fiords? (*If you do not know, read page 41 again.*) (6) What links can you see between the pattern of rivers, hill-ridges and rias in the south of Ireland?

As well as the geological links between Ireland and the rest of Highland Britain, the people of Ireland have much in common with those in Scotland, Wales and Cornwall. From an early date each of these districts was settled by Celtic-speaking peoples, who probably came from north-west France and the Rhineland. Celtic speech and customs have survived most strongly in Ireland, which was invaded by neither Romans nor Anglo-Saxons. One important development, however, was the conversion of the Irish to Christianity by St. Patrick between A.D. 400 and 450. A century after the Norman Conquest, invaders from England began to drive the Irish from the lowlands around Dublin. Later on large numbers of English and Scottish Protestants were settled in the ' Plantation of Ulster ' in Northern Ireland. Ulster is still largely Protestant, whereas the rest of Ireland is 94% Roman Catholic.

Although Great Britain and Ireland became a ' United Kingdom ' in A.D. 1800, the Catholic Irish were never content under English rule, and after years of bitter struggle the island was separated into two parts in 1921. The political details of the division are shown on the map on page 226.

* The whole Irish rail system is in danger of closure because of lack of custom.

224

Because of her rich green pastures, in which grasses, ferns and mosses flourish throughout the year, Ireland is commonly called the 'Emerald Isle'. (7) With the help of this map and the climate charts (and remembering what we have just learnt about the geology of Ireland) can you explain why:—

1. Grass can grow at all seasons in Ireland; and
2. One-half of the entire land of Ireland is in permanent pasture?

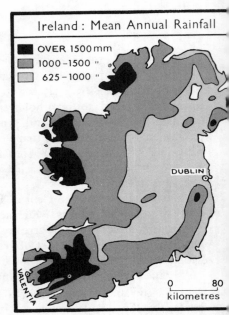

Ireland: Mean Annual Rainfall

■ OVER 1500 mm
■ 1000–1500 "
□ 625–1000 "

DUBLIN

VALENTIA

0   80
kilometres

(8) For both Valentia and Dublin work out (a) the total annual rainfall, and (b) the mean annual temperature range. (9) Can you explain the contrasts? (10) Is it true to say that the outstanding features of the climate of Ireland are its humidity and equability?

(11) Why is rainfall in Ireland neither so heavy along the west coast, nor so light along the east coast, as in England and Scotland? (Hint: look again at the relief map on page 63.) The Midland Plain in England gets an average rainfall of less than 30 inches. (12) How does this compare with the rainfall in the Central Lowland in Ireland? How do you explain the difference?

The district around Dublin is the only part of Ireland in which cereal crops can be ripened without difficulty ((13) Why?) and so it is not surprising that most Irish farmers concentrate on rearing animals—cattle, sheep and pigs. One crop is well suited to the damp, equable Irish climate—the potato. During the 18th century potato-growing expanded so much that it became the Irish people's chief food. Up to A.D. 1845 its cultivation made pos-

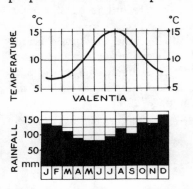

VALENTIA

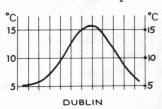

DUBLIN

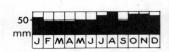

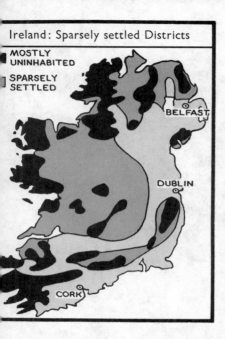

Ireland: Sparsely settled Districts

MOSTLY UNINHABITED

SPARSELY SETTLED

BELFAST

DUBLIN

CORK

sible the growth of population indicated on the chart below. Then came a disastrous blight which completely ruined the potato crops in 1846–1847. Many thousands died of starvation and the countryside was left poverty-stricken and desperately unhappy.

Within fifteen years after the famine $3\frac{1}{4}$ million emigrants poured out of Ireland —mostly to the U.S.A., to Canada or to other parts of Britain. A steady, but slower, emigration from Ireland has continued right up to the present time. (*14*) The descendants of these emigrants are estimated to number at least 16 million: how does this figure compare with the total population today?

Considerable parts of Ireland are either completely uninhabited or very sparsely settled. (*15*) Compare the map above with the one opposite, and with the relief map on the next page: what links can you trace between the three maps? There are no important coal or iron deposits in Ireland, so densely populated industrial conurbations like those on the English coalfields have never developed there. Most Irish people live in the country, in widely scattered farms or small market towns. There are only three large cities in all Ireland (see map above). (*16*) Each is referred to in the next few pages: meanwhile can you suggest why all of them should have grown up on the east coast?

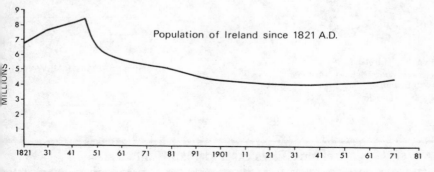

Population of Ireland since 1821 A.D.

MILLIONS.

9 8 7 6 5 4 3 2 1

1821 31 41 51 61 71 81 91 1901 11 21 31 41 51 61 71 81

226

Northern Ireland
Forms part of the United Kingdom of Great Britain and Northern Ireland.
Capital—Belfast.

Giant's Causeway

North Channel

DONEGAL

Foyle

LONDONDERRY
BALLYMENA
SPERRIN MTS.

ANTRIM PLATEAU

LARNE

Bann

Antrim

BELFAST

L. Erne

NORTHERN IRELAND

Lough Neagh

CRAIGAVON

DOWN

Slieve Donard

MOURNE MTS.

SLIGO

DUNDALK

MAYO

WESTPORT

I R E L A N D

Boyne

CONNEMARA

ATHLONE

DUBLIN

DUN LAOGHAIRE

GALWAY

Shannon

Bog of Allen

Barrow

WICKLOW MTS.

Power Station

LIMERICK

Golden Vale

WEXFORD

ROSSLARE

TRALEE

MUNSTER

Suir

WATERFORD

MALLOW

Valentia Island

Lee

CORK

COBH

R. Blackwater

LAND OVER 200 m

- - - - MAIN RAILWAYS

0   40   80
kilometres

Republic of Ireland (Eire)
Has no political links whatever with Britain.
Capital—Dublin.

Ireland : Physical and Main Railways

NORTHERN IRELAND

Look again at the map on page 222—it shows that Northern Ireland may be regarded geologically as an extension of Scotland. Today the two countries are separated by the sunken glen of the North Channel, but the faults which run across Scotland continue into Northern Ireland, where they converge and gradually die out in the Lough Erne district.   Metamorphic rocks like those of the Scottish Highlands reappear in the Sperrin Mountains, and the Southern Uplands are continued in the ancient hills of Armagh and the Mourne Mountains.   The area between the faults makes an extension of the Scottish Lowlands, but unfortunately for Northern Ireland there are no important coal seams in the Carboniferous rocks here.

This simple threefold division is obscured, however, by overlying masses of glacial drift, and also by two great upwellings of molten magma (page 7).   One of these, an outpouring of basalt lava, forms the area around and including the Antrim Plateau. Cooling quickly and crystallising into six-sided columns, the basalt has given rise, at the 'Giant's Causeway', to the spectacular cliff scenery sketched below (*and see p. 17*).

In the south of County Down, by contrast, the Mourne Mountains are the exposed remains of a deep-seated, slowly cooled flow of granite.   (*17*) What similar hills are found in the Lowlands of Scotland?   (*18*) Why is volcanic activity often associated with the foundation of rift-valleys?

From Ulster it is easy to imagine the time when Ireland was linked by land to the rest of Britain.   " One can look across the North Channel far into the complexity of lochs and firths and islands and see houses and fields over there; and one can go over in a small open boat.   In the same way, from County Down one can see the Isle of Man with the Cumberland and Westmorland hills beyond it.   And an ascent of Slieve Donard leaves one free to boast of having seen Wales.   It is true that there may be some family controversy about that dim shape on the south-eastern horizon; but that is where Wales ought to be, so surely we must have seen it."[1]

[1] *The Times*, Northern Ireland Supplement.

NORTHERN IRELAND: FARM FACTS

Farm businesses*	31 715
(a) 1 to 20 ha.	11 448
(b) over 20 ha.	14 267

Average area of crops and grass for farm businesses = 25 ha

Total agricultural labour force (excl. wives of owners)	65 700
Tractors in 1939	858
Tractors in 1975	38 748

AGRICULTURAL PRODUCTION

	£m.
Total	416·4
Cattle	154·9
Milk	99·9
Pigs	51·2
Eggs	34·1
Potatoes	24·7
Poultry	22·2
Horticulture	11·5
Sheep	8·7
Cereals	5·5
Wool	1·2
Other items	2·6

AGRICULTURAL EXPORTS

	£m.
Total food and drink (about one fifth of all exports)	314·3

The main agricultural exports, in order of importance, are:

Meat; Eggs and poultry; Bacon, ham and live pigs; Live cattle; Milk and milk products.

Over 90% of these exports go to Britain.

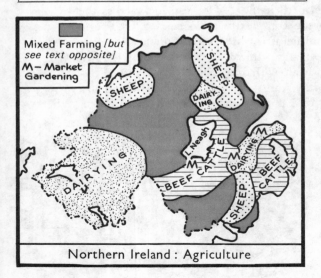

Mixed Farming [but see text opposite]

M—Market Gardening

Northern Ireland : Agriculture

This view of the little houses, farmsteads and fields of County Down, in the shadow of the Mourne Mountains, is characteristic of much of Northern Ireland. A century ago most of the land lay in large estates, owned by a few great families, and was rented out in small-holdings. "The countryside is a colonial one, with hardly a building that is really ancient, and with little road-junction towns rather than villages of the English kind. The landscape still presents its pleasant dappled scene, with tiny fields and tiny farms and fat white gateposts—a scene formed by the cutting up of the land into very small tenant holdings before the Irish Land Acts converted tenants into proprietors."[1]

[1] *The Times*, Northern Ireland Supplement.

Northern Ireland is still a country of small family farms. Through time, however, many of the smallest holdings have been combined to form larger, more profitable units. This has been partly the cause and partly the result of men leaving the land to work in industry; their place has been taken by machines. Thus the average size of farms has risen, and in the 1960s alone it doubled. Even so, more than half of the farms are still worked by one man, often as a spare-time occupation. A major problem is the cost of shipping food exports to the U.K., which adds 10–20% to their price and makes them less competitive.

Despite the scene shown above, crops are of minor importance and are mainly grown for the farmer's own use. After hay, barley is easily the leading crop, with oats and potatoes a poor third and fourth respectively. Study the details given opposite and then:—

(*19*) Attempt an explanation of the map, after comparing it with the relief and rainfall maps on pages 224 and 226.

(*20*) Calculate the average number of workers per farm.

(*21*) Draw a ' bar ' diagram to show the relative value of the various items of agricultural production.

(*22*) Draw a similar diagram for the agricultural items exported.

(*23*) What percentage of (*a*) foodstuffs and drink, and (*b*) livestock, exported from Northern Ireland goes to Great Britain?

In spite of her lack of native coal and minerals Northern Ireland has become an important manufacturing region. The chief industrial centre is Belfast, which with 392 000 inhabitants contains nearly one-third of the country's population. The map on page 226 shows:—

1. The port's sheltered position at the head of Belfast Lough;
2. The various 'lowland corridors' converging on the city from its hinterland (which now includes all Northern Ireland).

Imports of coal, iron and steel from central Scotland and Cumbria were used in building up Belfast's industries, the chief of which are shipbuilding, engineering and textiles. Water transport is nearly ten times cheaper than carrying goods by rail, and the cheapness and ease with which industrial raw materials can be imported have so favoured Belfast that it now has Britain's third largest shipbuilding industry. (*24*) (*Where are the two larger?*) The Lough is easily dredged and its wide waters have facilitated the launching of large vessels. The principal shipbuilding firm—Harland and Wolff—have ten slipways and a great marine engineering works in Belfast. Below is a picture of some of the construction yards. Many naval craft, tankers and cargo liners are also built or refitted in the Belfast shipyards.

These modernised and re-equipped Belfast shipyards are the most up-to-date in Britain. They include the world's largest dry-dock in which bulk carriers up to 1 m. tonnes can be built and then floated out. Other products include LPG carriers and ferries.

Aircraft manufacture forms an important part of the engineering industry in Northern Ireland. Britain's first aircraft designed for vertical take-off and landing was produced by the Belfast firm of Short Brothers & Harland. Nine thousand people are employed in Short's factories from which come a succession of aircraft, guided missiles and other aeronautical and non-aviation products. Many goods are exported, including light air-freighters and 30-seater 'commuter jets', which go mainly to the U.S.A. Component exports include doors for U.S. Jumbo jets and wings for Dutch Fokker planes.

General engineering includes making agricultural and textile machinery and has greatly expanded in recent years: new products include:—

Tape recorder, radio and television equipment; precision instruments; computers; vacuum cleaners; automatic-control systems.

Northern Ireland has been important for its linen manufacture since the 13th century. For long the cloth was made by peasants in their own cottages, but power looms were introduced in the mid-19th century and production now takes place in large factories. The raw material, flax, was formerly all produced locally, for the well drained clay soils and damp climate of the Ulster lowlands favoured its growth. Flax is not now grown in any quantity, and the linen industry depends very largely on imports, largely from Belgium and the Netherlands.

Belfast is the main home of the linen industry and there are mills in most towns throughout the country, but to meet changes in

Although linen cloth is a very high-quality fabric, it is difficult and costly to produce. For long the Irish linen trade has been contracting and losing ground to cheaper textiles. Export sales (chiefly to North America and Australia) are maintained by concentrating on quality fashion fabrics like this summer dress in white linen.

The photograph shows part of a nylon factory in County Antrim. In every way, includi... the ' man-made ' fibre and the large-scale methods of production, such a factory is very f removed from the traditional world of hand-pulled, hand-spun flax and hand-woven line... yet it is factories like this that more truly mirror the Northern Ireland of today. Most of t... cloth woven in Ulster is finished there, for the damp climate and soft water are very suitab... for bleaching, dyeing and printing textile fabrics.

modern tastes and fashions the textile industry in Northern Ireland now processes many other fibres besides flax. A small, but ancient, woollen industry is expanding with the production of ' Acrilan ' fibre in a new factory at Coleraine. Recent developments include the use of rayon in the old-established shirt, collar and pyjama factories in Londonderry, and the production of cotton, rayon, ' Fibro ' and glass filament.

For many years unemployment has been a serious problem in Northern Ireland, chiefly due to a gradual contraction in linen and agriculture. To meet the difficulty, attempts are being made to foster new types of production, such as the expansion in textiles just mentioned, and a Development Programme (*see opposite*), underpinned by massive economic support from the U.K., was launched in 1970. Although progress in broadening the economy has been made, Northern Ireland has certain disadvantages as a location for manufacturing. These include:—

1. The need to import most raw materials and fuel; 2. The need to export much finished produce, as the local Irish market is too

small; 3. The high transport and handling costs for industrial goods; 4. The political and social unrest.

New services designed to reduce carrying costs and travel time should help to attract industry to Ulster. These include:— 1. A transport ferry service with vessels (like that shown above) specially built to carry ro–ro vehicles and containers; 2. Improvements to docking and handling facilities at the ferry ports (see page 238); 3. Improved air services for passengers and freight.

The Northern Ireland Development Pro-mme is intended to rectify faults in the onomy of Ulster, where the average ome is only 75% that of the U.K. e two main problem areas are: (i) parts central Belfast, which are congested d decayed, and (ii) the mainly Catholic al areas in the south and south-west the Province, where farm incomes are y low and many men are unemployed. e main objects of the Programme are:—
1. To limit the growth of Belfast by offering resettlement grants to those firms and residents who agree to move out.
2. To establish New Towns at Londonderry, Antrim-Ballymena, and Craigavon (map p. 226).

3. To set up small workshops and factories in rural settlements.

To date, most progress has been made at Craigavon, which has 80 new factories, and in Antrim-Ballymena, where Michelin and British Enkalon are important newcomers. Generous financial help is given to outside firms which come to Northern Ireland, and this has attracted many American, Dutch and German manufacturers in processed foods, electronics, engineering and textiles. Over 30 U.S. textile firms, for example, have located in the Londonderry district, which now has Europe's largest concentration of man-made fibre factories.

THE REPUBLIC OF IRELAND

WESTERN IRELAND includes some of the most remote and desolate countryside in the British Isles. The few inhabitants of this barren, rock-strewn land lead a hard life of peasant farming (coupled with fishing on the coast) similar to that of the Scottish crofters (see page 69). High-quality hand-knitting and hand-weaving are traditional 'cottage industries' in Donegal, though the famous Donegal tweed now comes mainly from small factories.

Tourists visit the region for its beautiful lake, coast and mountain scenery. There are no large towns, but some small market centres like Sligo and Galway. These ports have good natural harbours, but little trade because their hinterland is so unproductive.

SOUTH-WEST IRELAND is similar to the region just described. It is mountainous, remote and in places barren, but, unlike the north-west, has many broad valleys. (*25*) Look at the maps opposite: what rocks form (*a*) the hill ridges; (*b*) the valleys? (*26*) Can you suggest why the relief and geology are related in this way? (*27*) Why, too, do rivers such as the Bandon, Lee and Blackwater all follow right-angled bends to the south before entering the sea? (Compare the Wealden drainage, pp. 191–3.)

The valleys are floored with alluvium and boulder clay, and are very fertile. Towards the east they are also sheltered from the full force of westerly gales and the mild, damp climate is very suitable for growing grass. Sugar beet also does well, but little grain is grown. ((*28*) Why not?) Hence the fertile lowlands, especially in the 'Golden Vale' of Tipperary and Limerick, are famous for dairying. Most farmers here send their milk to

A typical landscape in Donegal: notice the small-holdings (twelve homesteads are visi in this photograph alone) into which the stony land is divided. Soils are thin and infertile, a the holdings so small that living standards are very low. Rainfall is very heavy, but so hay, oats and potatoes are grown on little arable patches, whilst sheep and some cat browse on the rough hill pastures. Most farmers also keep a few pigs and poultry. Plans improve conditions in rural Ireland are given on p. 237.

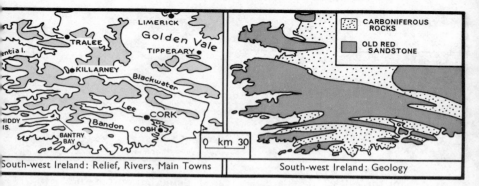

CARBONIFEROUS ROCKS

OLD RED SANDSTONE

0 km 30

South-west Ireland: Relief, Rivers, Main Towns | South-west Ireland: Geology

co-operative creameries where it is made into cream, butter and cheese for sale in Ireland and for export. The skimmed milk is returned to the farms, where it is fed to large numbers of pigs. Bacon and ham are cured in factories at Limerick and Cork. In addition to dairying, south-west Ireland has an expanding tourist industry, for the ria coastline and the mountainous district around Killarney include some of the finest scenery in the country.

The coastal scenery has already been marred in places by industrial building. In Bantry Bay, e.g., a petroleum terminal has been built; at a deep water berth off Whiddy Island (*map A*) huge supertankers discharge their cargoes of Middle East oil either directly into smaller ships or into storage tanks. The oil is then taken to various European ports to be refined.

Cork (132 000), the second largest town and second port of Ireland, is the main centre for the export of live cattle and dairy produce from the Golden Vale. It also has an extensive and growing group of industries, details of which are given on the right. Cruise liners call at the 'out-port' of Cobh, and the airport has a busy freight traffic.

> oil refining
> vehicle building
> engineering
> flour milling
> shipbuilding
>
> Manufacture of:—
> steel, chemicals,
> clothing, tyres.

SOUTH-EAST IRELAND is the only part of Ireland in which cereal crops are grown on a large scale. This is because the district has higher summer temperatures and less rainfall than elsewhere. ((*29*) Why?) Barley and oats are the main crops, and some wheat is grown along the east coast in County Wexford.

Farther west, beef-cattle rearing is very important, especially in the fertile river valleys of the Nore, Barrow and Suir. Large quantities of root crops, chiefly turnips and mangolds, are grown as fodder

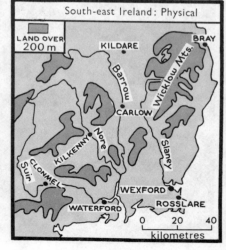

South-east Ireland: Physical

LAND OVER 200 m

0   20   40
kilometres

THE CENTRAL LOWLAND
showing the position of
DUBLIN

for the cattle. Pigs are also reared, being fed on potatoes. ((30) Where in England are the most important districts for (a) barley and wheat (pp. 242–5); (b) beef-cattle (p. 251)? (31) What similar geographical conditions apply there?)

The Wicklow Mountains form the largest mass of granite in the British Isles. The barren moorlands which cover them support very few people, but large numbers of sheep. In the east the mountains are dissected by beautiful wooded vales, like Glendalough, where arable and cattle farming is possible. (32) On the narrow coast plain around Bray, market gardening and dairying is important: to what large near by city do you think the produce is despatched?

THE CENTRAL LOWLAND, is in many places very badly drained. (*See photo and p. 223.*) Farms and villages are scattered over the area, mostly on patches of glacial gravels. On rich, wet pastures away from the bogs an enormous number of beef-cattle are reared for export to England, so that this part of

Ireland may aptly be called 'beef and bog' country. Dairying is unimportant, for transport services are so poor that no regular collection of milk is possible. Neither do the wet, marshy soils favour arable farming: hay is the chief crop, though some oats, barley, potatoes and turnips are grown as additional cattle food.

To the west of the Shannon the underlying limestone appears at the surface, and on its better-drained pastures large numbers of sheep are reared.

Patches of bog cover much of the Central Lowlands, where for centuries farmers have cut and dried peat for use as a fuel. Now peat is also used to fire seven electricity power stations.

Notice in the map opposite how communications converge on Dublin, the natural outlet for the products of the Central Lowland. Dublin is Ireland's chief port and the list of its exports (overleaf) shows that its hinterland includes the whole of the country. (33) Where are the main sources of each of these products? (34) What geographical explanation can you give for the lists of Dublin's manufactures and imports?

PROGRESS AND PROBLEMS IN RURAL IRELAND

Ireland's 180 000 farmers produce enough meat and dairy produce to supply the home market of 3 million, and a further 4 million overseas. About 80% of the total farmland is under grass, which is by far the most important crop. Heavily fertilised with nitrates it supports no fewer than 7 million cattle, most of them Friesian dairy cows, or Hereford beef stock. One-half of all farms produce milk, 84% of which is processed for export, mostly to the U.K. Since Ireland joined the Common Market, however, agricultural exports to the Continent have increased by 60%, and the high EEC prices for milk has helped raise farm incomes by at least one-third.

The long growing season (see p. 204) allows cattle to feed out of doors two months earlier and one month later than in most other Common Market countries. ((35) Where in the U.K. are there similar climatic conditions?) By reducing the amount of money needed for indoor feedstuffs this makes Irish milk the cheapest to produce in Europe. Grass is also the main feed for the 2.18 million Irish beef cattle which are marketed each year. Of these one-half are slaughtered and processed (mostly for export), and the rest are exported live.

Despite recent improvements, Irish farm yields are relatively poor—even milk production (2520 litres per cow) is 25% below the EEC average. Many farmers in the remote west and south-west find it hard to make a satisfactory living: re-read pages 224–5 and the last four pages and (36) make a list of the physical handicaps under which they labour. In fact the *lack of opportunities* for young people in central and western districts of Ireland is one of the country's major economic and social problems. During the past 10 years well over 200 000 people, mostly young men and women from rural districts, have left the country.

Other problems stem from *the very small size of Irish farms*; about one-half are under 12 ha, and no less than 68 000 are smaller than 6 ha. Incomes from these small holdings are very low indeed. " A typical small farmer in the west has an income of c. £250 from farming (including the value of his own farm produce consumed by himself and his family), eked out by unemployment assistance and emigrants' remittances." (*The Times*.) Even so, owners of these tiny plots fight to retain their land because of the prestige which land ownership confers.

To raise rural earnings the Government aims to establish large numbers of small light industries in country districts so that small-holders can have additional part-time employment. Partial industrialisation of this nature would suit the Irish economy because there is not enough large-scale factory growth as yet to give full employment to migrants off the land. A recent Government attempt to form larger farms by buying out and pensioning off small-holders at generous rates was abandoned because of almost total lack of support.

Matters are made more difficult by the population 'explosion' of recent years. Ireland has the highest birth-rate in W. Europe. The present population of 3.192 million is likely to reach 3.5 million by 1985, and 4.3 million by the end of the century.

Industries	Imports	Exports
Clothing, Electrical goods (cables, etc.), Foodstuffs, Beer, Ship repairing, Railway engineering	Coal, Cattle-cake, Fertilisers, Maize, Wheat, Manufactured goods	Live cattle, Sheep, Pigs, Dairy Produce, Bacon and ham, Beer and spirits

Compare Dublin's position with that of Limerick (see map opposite). Although Limerick has a good harbour it remained unimportant until quite recently, for it faces away from those countries which import produce from the Central Lowland. Despite its remote position, however, Limerick has gained fame, for the first major source of hydro-electric power in Ireland was developed nearby on the Shannon. Electricity from the Shannon and other power stations is sent by 'grid' cables to all parts of Ireland. As there are no large workable deposits of coal this power is extremely valuable. It has brought modern lighting and electrical machinery to scattered villages and farms, and has encouraged the growth of light industry. We have seen that the latter mostly involves the processing of agricultural products, in creameries, breweries, distilleries, beet-sugar refineries and bacon factories. To meet an increasing demand for power, especially from factories in Dublin and Cork, seven peat-fired power stations have been built. Today roughly one third of Ireland's power comes from peat and water and the rest from coal and petroleum.

Limerick is also benefiting from its proximity to Shannon International Airport (see map), where a Free Industrial Estate was

Main Ferry Routes to Ireland and Rail Links

This map shows that Dublin is well placed for trading connections with Liverpool and the English Midlands. There is also a fast ferry service from Dun Laoghaire (its out-port) to Holyhead, with express train links to London. Dublin Airport is one of the largest in Europe, handling more passengers than the ferry boats, as well as transporting cars and other goods to Great Britain and the Continent

created in 1958. This Estate is a free-trade zone, i.e. no import or export taxes or duties are paid by firms within its boundaries. Many foreign firms have set up factories at Shannon, attracted by the very favourable trade conditions offered

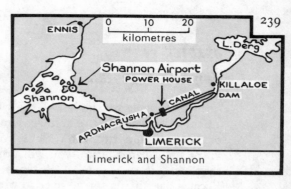

Limerick and Shannon

by the Estate: the main advantages are (i) no income or company taxes are paid on profits from exports for twenty-five years; (ii) cash grants are available to build and equip factories; (iii) government-sponsored loans can be obtained at favourable rates of interest; (iv) local labour supplies are plentiful and cheap; (v) finished goods can easily be exported, either by air from Shannon or by sea from Limerick. The Shannon Development Corporation* has attracted various light industries, including Japanese radio, Dutch piano and British knitwear firms.

* This Corporation is controlled by the ish Development Authority (I.D.A.), a overnment agency which aims to attract eign firms and capital to all parts of the public. Altogether over 700 firms, giving 000 new jobs, have been established th foreign help during the past 15 years. tal-goods, processed foods, and garments e typical products. The main industrial wth areas are (i) Dublin–Dundalk; (ii) annon–Limerick–Ennis and (iii) Cork rbour. Galway, Sligo and Tralee are the st notable of many smaller centres. ') Show this information by adding three ge and three small labelled dots on an line sketch-map of Ireland. Give your shed map an appropriate title.

The I.D.A. especially seeks industries ich will (a) grow quickly, (b) increase land's export trade and (c) employ large mbers of men. As a location for expand- foreign firms Ireland's distinct ad- tages include a plentiful supply of labour d cheap land, access to deep water ports, undant water and clean air. Japanese, nadian and U.S. firms are especially keen locate in Ireland in order to gain tax-free ess to the Common Market. About one- f of all industrial investment in Ireland is w derived from the U.S.A.

The I.D.A. also maintains industrial estates at Galway and Waterford, runs industrial training schools and has a ' Science Research Park ' at Naas (map, p. 237). Other industrial estates are run by private firms at Dublin, Cork and Castlebar. New turf- and oil-fired power stations have greatly increased power supplies. The result of all these incentives is that Irish industrial output is expanding by 7% p.a. The Irish economy is vitally dependent upon its growing exports of manufactured goods. (38) Draw a diagram to show that these increased their share of total exports from 17% to 51% in 1976. Important exports include textiles and apparel, chemicals, machinery and precision engineering goods, metals and metal concentrates. The latter follow the discovery in 1970 of immense deposits of zinc and lead at Navan, Co. Meath. An increasing amount of invisible exports come from Ireland's growing tourist industry, which now employs 10% of the country's workforce (the largest single employer after agriculture).

Despite an impressive record, the I.D.A. must provide a total of 420 000 new jobs by 1986, i.e. almost 50% more than the existing workforce, if it is to cure unemployment.

CHAPTER 20

Farming in Britain

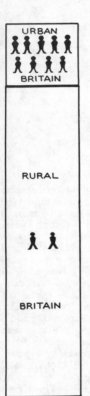

URBAN BRITAIN

RURAL BRITAIN

𝄇 = 5 Million People

THE DIAGRAM on the left shows what a very large proportion of British people live in towns and cities. Unless you are one of the small minority who really live in the country—in a hamlet, village or isolated farm—you will probably be surprised to know that agriculture is one of Britain's most important industries. Because farming is spread over so wide an area, we tend to forget that one in every thirty-four workers is employed in agriculture. (*1*) Work out from the diagram below what proportion of the surface of Britain is farmland.

Most British farmers grow a variety of crops and also rear animals—i.e. they practise **mixed farming**. They also get a very high yield per hectare from their farms by using scientific methods of crop growing and animal rearing. These include selecting the best seeds, breeding high-quality animals and keeping a strict check on pests and diseases. ((*2*) Find out what precautions are taken when foot and mouth disease breaks out on a farm.) Soils are kept fertile by adding manure and fertilisers, and by the rotation of crops (see pp. 164–5). The productivity of British farms is increasing, too, as farmers make greater use of apparatus such as combine harvesters, milking machines and tractors. Yet in spite of such **intensive farming** we produce in Britain only about one-half of the food supplies we need.

Although mixed farming is still typical of much British agriculture, farmers in some areas have more pasture than ploughed land; elsewhere they concentrate on growing crops, and rear few animals; while some districts are particularly noted for one or two particular crops, such as fruit, hops, potatoes or flowers. In the course of this book we have noted many such areas of *specialised farming*.

What factors influence the distribution of these various farming types? In general, a farmer's choice is affected by the nature of the *soil*, *climate*, *relief* and *drainage* of his land. Modern *fast transport*, too, encourages him to specialise

LAND USE IN BRITAIN

Urban areas Airfields Factories Roads, etc.	Forest and Woodland	Rough grazing (mostly in Scotland)	Grain crops	Other crops (*mainly temporary grassland*)	Permanent grassland

Before it was cleared for agriculture this typical English arable farmland was completely hidden by deciduous trees like those in the picture. (3) Can you suggest why these few trees have been left standing? An average of 21 000 hectares of farmland, mostly like that shown here, are lost to farming every year. (4) How?

because he is not limited, as were his ancestors, to a local market and can therefore sell anywhere whatever produce he is best placed to grow or rear. Often, however, his year-to-year plans are shaped by *E.E.C.* (Common Market) *decisions* about the kind of crops and livestock that Europe's farmers should be encouraged to produce.

The main areas of arable land in the British Isles are shown on this map. (5) What links can you see between it and (a) the relief and rainfall maps on pages 62 and 63; (b) the isotherm maps on page 52; and (c) the geological map on pages 64–65?

(6) Shade and mark a sketch-map of Eastern England to show East Anglia, Lincolnshire, the Vale of York and East Yorkshire. (7) Read again what has been said about these districts (use the index and table of contents) and make a list showing what advantages they have in common. (Hint: relief? climate? soils?)

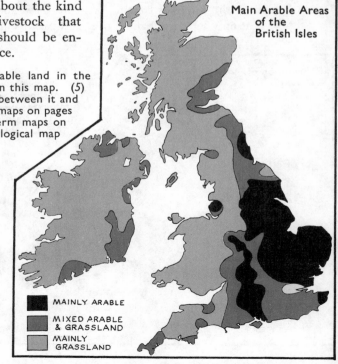

Main Arable Areas of the British Isles

■ MAINLY ARABLE

■ MIXED ARABLE & GRASSLAND

□ MAINLY GRASSLAND

242

WHEAT grows best in Britain where there is:—
  less than 750 mm mean annual rainfall;
  some rain when the grain is in its early stages of growth;
  warm, dry sunny weather for ripening and harvesting;
  land flat enough to allow the use of big machines;
  fertile and fairly heavy but well-drained soil.

By comparing the map below with that on the previous page we see that the chief areas of wheat-growing in Britain largely coincide with the richest arable districts, for it is there that climate and soil conditions most often match those listed above.

Because bread is a staple food the government encourages farmers to grow as much wheat as possible by guaranteeing them a minimum price for all they send to market. Farmers thus find it profitable to grow wheat wherever conditions suit the crop. By intensive methods of cultivation they get a high yield (*see table*), but the total crop, though nearly doubled in the past thirty years, is still less than half of Britain's requirements. Only about half is 'hard' wheat of the type used for flour-milling; the rest is fed to livestock, mainly poultry.

Wheat is widely distributed throughout the British Isles, yet as a main crop it is very localised. The dotted areas on the map show where it is of (purely local) importance in Scotland and Ireland; but in Norfolk or Suffolk alone more wheat is grown than in the whole of Ireland, or in Scotland and Wales combined.

Not only wheat but milk, beef and various other farm products

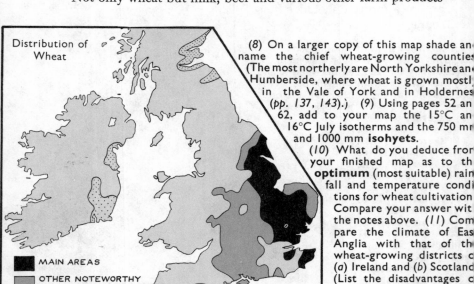

Distribution of Wheat

MAIN AREAS
OTHER NOTEWORTHY AREAS
SEE TEXT

(8) On a larger copy of this map shade and name the chief wheat-growing counties (The most northerly are North Yorkshire and Humberside, where wheat is grown mostly in the Vale of York and in Holderness (*pp. 137, 143*).) (9) Using pages 52 and 62, add to your map the 15°C and 16°C July isotherms and the 750 mm and 1000 mm **isohyets**. (10) What do you deduce from your finished map as to the **optimum** (most suitable) rainfall and temperature conditions for wheat cultivation? Compare your answer with the notes above. (11) Compare the climate of East Anglia with that of the wheat-growing districts of (a) Ireland and (b) Scotland. (List the disadvantages of Cornwall for wheat-growing. Do not confine your comments to climate.)

Combine harvesters at work in Wiltshire. Each of these machines can cut, thresh and load half a hectare of wheat in an hour. It is economical to use them only where there are large fields, unhindered by hedges or steep slopes.

Here, on Salisbury Plain, is one such area where ' prairie farming ' ((14) Explain the point of this term) has become common. The level or gently rolling countryside of East Anglia is another. Fields of 20 hectares are normal on farms like these. How big this is can be better realised if we know that half of the 250 000 farms in England and Wales have a total area of less than 20 ha.

are sold at, or above, prices guaranteed by the government in accordance with E.E.C. agreements. The trading losses, if any, are paid for by the tax-payers—your parents, and all other citizens of the Common Market countries. (*12*) Do you think this is a sensible plan? What reasons for or against it can you suggest?

Modern wheat-farming has meant not only bigger, heavier machines and the removal of thousands of kilometres of hedge-rows. It also involves massive doses of chemical fertilisers, weed-killers and pesticides to replace traditional ' mixed farming ' practices. Some scientists say the results have harmed wildlife and the soil itself. (*13*) What do you know about the case for and against modern farming methods?

WHEAT YIELDS (tonnes per hectare)	
Britain	3·9
Canada	1·8
Australia	1·5
Argentina	1·2

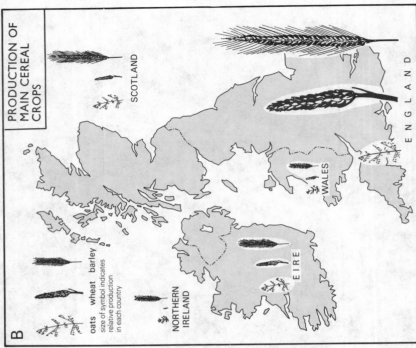

B — PRODUCTION OF MAIN CEREAL CROPS

SCOTLAND

oats wheat barley
size of symbol indicates
relative production
in each country

NORTHERN
IRELAND

EIRE

WALES

E N G L A N D

(16) State in your own words what map (B) tells us about cereal growing in Britain, and suggest reasons for the contrasts shown.

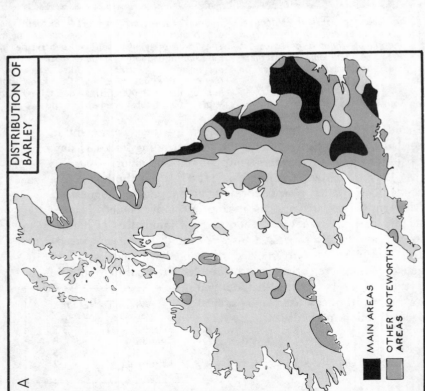

A — DISTRIBUTION OF BARLEY

MAIN AREAS

OTHER NOTEWORTHY
AREAS

(15) Using your atlas, identify by suitable names the areas shown on map (A) as being chiefly important for barley growing.

BARLEY...

does best in practically the same conditions as wheat;

does better than wheat on dry, sandy 'light' soils;

can ripen at lower mean temperatures than wheat;

has been scientifically developed into special types which resist disease better than wheat, and can therefore be grown in rainier districts;

was used in Britain, until about 1950, chiefly for 'malting', i.e. for making beer, whisky and vinegar;

has expanded enormously in area since 1950, and is easily Britain's leading crop (except for grass);

is now used chiefly for fattening pigs and (to some extent) beef cattle.

OATS...

grows best in areas suited to wheat and barley;

also thrives under moist, cool conditions;

until about 1950 was the only important cereal crop of the wet, western and cool, northern parts of Britain;

was (and is) used mainly as fodder for working horses; also (decreasingly) for human consumption;

has been largely replaced, even in Scotland, by the new, hardy types of barley.

BRITAIN—MAIN CEREAL CROPS
(area sown, in thousands of hectares)

	WHEAT	BARLEY	OATS
1938	803	412	998
1975	1035	2345	233

Apart from wheat, the main cereal crops of the British Isles are barley and oats. In the past forty years their relative importance has drastically altered, as illustrated by the table above. Almost half the British oat crop is grown in Scotland and Wales, as compared with only one sixth of the barley and a mere 4% of the wheat.

Important facts such as these are the 'raw material' of geography, but make rather dull reading. It is much more interesting to show them in diagram form; as, for example, in map (B) opposite, which shows in more detail the facts mentioned in the previous paragraph. (*17*) Draw a similar diagram to illustrate the table above, using a small outline map for each year. (*Hint: to get the size of the 'ears' in correct proportion, draw them inside rectangles on squared paper and let each small square stand for (say) 10 000 ha.*)

(*18*) Design diagrams to illustrate the table below in a striking way. (*19*) Note, and try to explain, the contrasts between 1913 and 1975.

	OATS (thousands of hectares)	WORKING HORSES	FARMERS and FARM WORKERS	TRACTORS IN USE
1913	1040	1 330 000	1 543 000	hardly any
1975	233	10 000?	504 000	530 000

POTATOES are . . .

very hardy, needing little sun or moisture ;

grown throughout lowland Britain ;

best grown in deep, rich, loamy, stone-free soils ;

easily damaged in transit (few are imported); and so

grown for marketing in Great Britain ;

a staple British food, and good animal fodder ;

used to make spirits and industrial alcohol ;

grown especially in Ireland, mainly by small farmers, as pig-feed and for home consumption.

grown in Scotland for human consumption, and also very largely for ' seed ' (p. 77).

The map below shows, for England and Scotland, the main areas where potatoes are grown on a large scale for urban markets. (20) Identify by suitable names each of these areas, and name the large urban market or markets near-by to which the potatoes are sent.   In addition to the producing districts shown, early potatoes are grown in Cornwall (see opposite) and the Channel Islands.

SUGAR-BEET prefers a deep, fertile, stoneless soil—hence it grows well in areas where large-scale potato cultivation also takes place. Unlike potatoes, however, sugar-beet cannot stand a heavy rainfall, and needs a dry harvesting period in the autumn.   (21) To what extent is this contrast in rainfall and sunshine requirements reflected in the map?   (22) Where are the main districts of sugar-beet production?

Sugar-beet is a valuable crop apart from the sugar it yields. The tops, and the pulp from the factories, are fed to cattle. In eastern England sugar-beet has largely replaced turnips and mangolds as a root crop in arable rotation farming.

Some other branches of crop farming use comparatively small areas of land, yet their products are important to our everyday life.

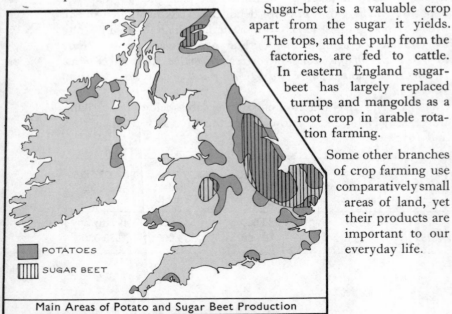

POTATOES

SUGAR BEET

Main Areas of Potato and Sugar Beet Production

VEGETABLES such as peas, carrots and brussels sprouts are mostly grown as field crops in rotation with cereals. The importance of climate and soil is shown by the fact that a quarter of all Britain's vegetables come from the Fens, and another quarter from East Anglia (*see pp. 165 and 167*).

HORTICULTURE is small-scale, intensive and highly-specialised crop farming. It is generally divided into the two types noted below. In both branches of horticulture British growers have lost much of their former advantage of being close to their market. Air freighters now deliver freshly-picked produce from France, the Mediterranean countries, even California, at prices which are still competitive because these crops cost so much less to grow in a warm climate. On the other hand the rise of the freezing and canning industries has lessened one of the British horticulturist's special hazards: i.e., a sudden spell of warm weather, producing a 'glut' of highly perishable produce and sending market prices down below the cost of production.

(*23*) Draw a sketch map to illustrate the facts given below.

(*24*) After studying a good atlas and the index references to earlier chapters in this book, suggest the factors favouring horticulture in each of the districts named, such as: *climate? soil? nearness to markets? shelter from cold winds?*

MARKET GARDENING . . .

is mainly concerned with fresh VEGETABLES (*e.g. early peas*) and SALAD crops (*e.g. lettuce, tomatoes*). CUT FLOWERS, BULBS and BEDDING PLANTS are other typical products. Market gardening demands little land, great care and knowledge and much labour, both skilled and unskilled. It is thus a most intensive type of farming, especially as regards the 2500 ha of crops grown ' under glass '.

Two factors have traditionally influenced the location of market gardens: (*a*) length of growing season, and (*b*) proximity to densely populated areas. ((*25*) *Explain this sentence.*)

Land near towns has now become very valuable, and transport is fast and efficient. ((*26*) *Suggest why market gardeners have, in the last few years, sold up and moved further out into the country.*)

Leading districts are: the Fens; the fringe areas all round Greater London; Cornwall and the Scillies (*esp. for spring flowers*); the Channel Islands (*spring flowers, early potatoes and tomatoes*); Bedfordshire; the West Sussex/Hampshire coastal fringe; the Vale of Evesham; south Lancashire; west Yorkshire; the Clyde Valley.

FRUIT GROWING . . .

is concerned either with ' orchard ' fruit (apples, plums, pears, cherries) or ' soft ' fruit (e.g. strawberries, blackcurrants). Both need shelter from wind and late frost, plenty of moisture and also plenty of sunshine. Sunshine is especially important for soft fruit.

British summers being so variable, fruit growing is a risky enterprise and is noticeably localised by climatic factors. It requires specialised skill and equipment and a reliable supply of seasonal labour for picking.

Leading districts are:

(*a*) *orchard fruit*: Kent (*over $\frac{1}{5}$ of all U.K. orchards*); East Anglia (*over $\frac{1}{6}$*); Vale of Evesham and lower Severn valley (*over $\frac{1}{7}$*); Somerset and Devon.

(*b*) *soft fruit*: the Fens and Kent (*each over $\frac{1}{5}$ U.K. total*); East Anglia; Vale of Evesham and lower Severn valley; south Hampshire; the Carse of Gowrie (*p. 77*).

SHEEP were England's chief source of wealth in medieval times, and much fertile land in the south-east was devoted to sheep rearing. Large quantities of wool were exported to Flanders by sheep farmers and merchants who made great fortunes (see page 166). Britain still ranks as a leading sheep-rearing country, but our population has increased so much that farmers today profit most by growing food crops and by rearing dairy and beef-cattle whenever their land permits.

Sheep are thus found mostly in districts unsuitable for crops or cattle, and especially on the mountain moorlands of the west and north-west. Look at the photograph on page 90: only sheep can feed well on rough hill-tops like these near Plynlimmon in mid-Wales, and even sheep can live here only in the summer. Each mountain sheep needs a half hectare, or more, over which to browze for young shoots of grass or heather—hence the rarity of boundary walls or fences.

Along the dry limestone and chalk escarpments, and in the arable lowlands of southern and eastern England, sheep have long been a traditional part of the rotation. Kale and other fodder crops are grown specially for them, and they in turn enrich the soil by manuring it and by treading it firm with their 'golden hoof' (*pp. 142, 164*). Yet since 1940 lowland sheep have become much less common. This has been particularly notice-able on the chalk downs, from which they have been driven out by the spread of more intensive farming practices.

The main sheep-rearing areas in the British Isles are labelled 1 to 10 on the map. (*27*) Identify each of them by suitable names. Hill

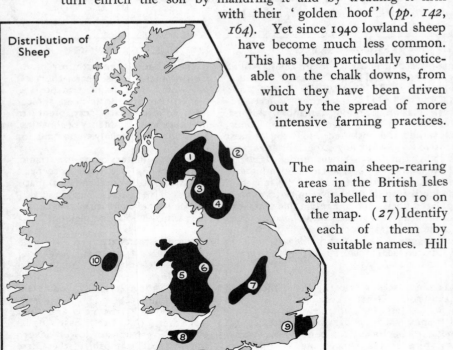

Distribution of Sheep

One of the sheep farmer's annual chores. What is happening here?

and lowland sheep are not completely distinct, despite the many (over thirty) different breeds, since thousands of hill sheep are sold each year to lowland farmers for fattening or cross-breeding.

On marshy ground sheep suffer from 'footrot' and other diseases. Rather surprisingly the greatest density of sheep anywhere in Britain is on Romney Marsh, in Sussex; but these alluvial pastures are well-drained.

Notwithstanding the large sheep population (about 28 million in Britain and 6 million in Ireland) we rely on imports for most of our wool and about half of our supplies of lamb. Total numbers vary from year to year, but have fallen slightly since the 1960s. Many lowland farmers use their land so intensively that they have no room for sheep; others find that experienced shepherds are rare, and increasingly hard to replace. Hill farmers and their wives are less inclined than they once were to accept the loneliness and hard conditions of life in the more remote districts; and many hill farms are being abandoned or sold for forestry.

Inside a milking shed near Saffron Walden, Essex. Using modern equipment, one man can milk dozens of cows in a very short time.

Both DAIRY CATTLE and BEEF CATTLE are fed largely on grass, or on similar plants such as clover and lucerne; and in the British Isles grass is easily the most widespread and important crop. Yet maps (A) and (B) show distinct contrasts in the distribution of the two types of cattle. Why?

Beef cattle are bullocks, bred and fattened for slaughter. Breeding is carried on anywhere where grass grows well and land is reasonably cheap; usually on hill farms or in fairly remote districts like Central Ireland. Most bullocks are sold, at some stage of their short lives, to grassland farmers nearer the ultimate market (e.g. in the English Midlands, p. 155), or to arable farmers (e.g. in N.E. Scotland or East Anglia) who fatten them on hay and roots. Not many farmers rear beef cattle as their main or only concern.

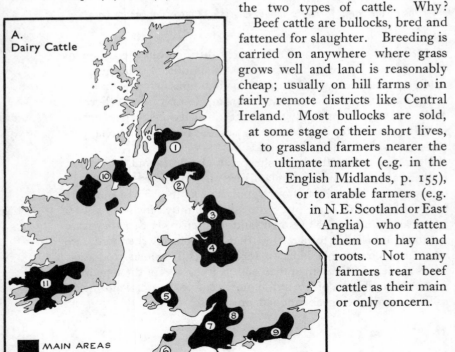

A. Dairy Cattle

■ MAIN AREAS

CHIEF DAIRYING DISTRICTS	CHIEF BEEF CATTLE DISTRICTS
Somerset and Dorset S.W. Ireland W. Scottish Lowlands and Ayrshire Coast Solway Firth Lowlands Wiltshire Lowlands of Devon and Cornwall N. Ireland Cheshire/Staffs. Plain Surrey, Sussex and S.E. Hampshire S.W. Wales Lancashire Plain	Irish Lowlands, esp. E. and S. Northumberland Lowlands and Vale of York Buchan Lowlands Solway Firth Lowlands S.W. Wales Severn Valley and Welsh Border Devon and Cornwall N. Wales S. Yorks and East Midlands

Dairying, by contrast, is a very specialised and demanding type of farming, and dairy cows are delicate animals. Equipment, extra feeding-stuffs and skilled labour seven days a week ((*28a*) *Why?*) are all expensive, and competition is keen. Most dairying, therefore, is in districts where climatic conditions help to keep down the cost of milk production. Most dairy farms, too, are 'family farms' of less than 40 ha. ((*28b*) *Why?* (*Hint: previous question.*))

The richest, juiciest grasses grow in the lowlands towards the west, where there is a well-distributed rainfall of over 750 mm per year, and where the milder winters allow cattle to feed outdoors for much of the year. The importance of these grasslands is reflected in both maps, but particularly in (A). Nevertheless (A) shows that dairying is fairly important in south-east England, where rainfall is lower and winters colder. These disadvantages are offset by the dampness of the clay soils and by the nearness of a huge market (*Where?*).

(*29*) Sort out the labels in the notes above to match the numbered districts on the maps. Then make separate lists of districts important for (*a*) both beef and dairy cattle; (*b*) dairy cattle only; (*c*) beef cattle only. (*30*) Explain (*b*) and (*c*) after revising the references in earlier chapters.

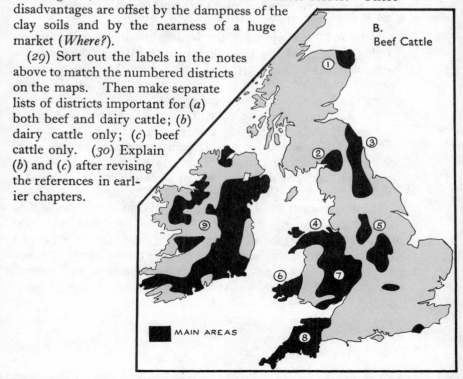

B.
Beef Cattle

MAIN AREAS

In winter the grass largely stops growing, and so in summer the farmer has to store away plenty of fodder to give to his cows when the pastures are bare, or when the weather is too bad for them to leave their stalls. The usual winter food for cattle is hay, and this machine on a Sussex dairy farm cuts and airs the grass so that it is ready to bale within twenty-four hours.

Milk and beef are easily the leading products of British and Irish farming, making up well over a third of total output (*see opposite*). As with all types of farming, methods have become very intensive. Controlled 'strip grazing' of pastures, using movable electric fences to avoid overgrazing or waste, has been replaced on some farms by 'zero grazing'; the animals live indoors and are fed on hay, mangolds, barley and manufactured rations.

Some cattle, like most pigs and poultry, are reared by even more intensive methods known as 'factory farming'. (*31*) What do you know about the pros and cons of this controversial topic?

There are about 8 million pigs in Britain, and some 140 million poultry of which 97% are fowls. Most fowls are kept in 'broiler houses' (for meat) or in 'batteries' (for eggs) and over half are in units of over 50 000 birds. By such methods a lot of 'farming' can be done in a very small space and in almost any district. Climate and soil are no longer important factors.

Potatoes 66 kg — Flour 77 kg — Milk 147 litres — Meat 60 kg — Other Vegetables 53 kg — Fruit 33 kg — Eggs 208 — Fish 6 kg

This drawing shows what each person in Britain consumes, on average, in one year. The source of certain of these foods is shown in the following notes.

SOURCES OF SOME MAIN FOODS CONSUMED IN GREAT BRITAIN (*Per Cent of Total*)

Commodity	Home-produced	Imported
Grain and Flour	60	40
Meat	78	22
Sugar	29	71
Potatoes	90	10
Fresh Milk	100	NIL
Butter	10	90
Cheese	60	40

(*32*) Draw a bar diagram 100mm long, divided proportionally, for each commodity mentioned. Discuss the results, and add comments.

(*33*) Then make bar diagrams to compare the value of those home-produced foodstuffs mentioned in the notes below: arrange the diagrams in their order of value:—

GREAT BRITAIN: AGRICULTURAL OUTPUT
(*% of total value of all farm produce sold. NB a large proportion of farm produce, such as hay, is not sold but is consumed on the farm.*)

Cereals	11·4	Pork and bacon	9·8
Potatoes	7·6	Milk and dairy produce	20·6
Sugar beet	1·6	Poultry and eggs	11·1
Beef and veal	17·1	Vegetables	6·6
Lamb	3·6	Fruit and flowers	3·6

(*34*) Finally, select from the above notes any TWO foodstuffs. For each one you select, (*a*) state TWO places in Great Britain where it is produced on a large scale; and (*b*) explain the reasons why it is so important a product in the districts you mention.

CHAPTER 21

Population, Energy Resources, Trade and Transport

(A) POPULATION

WE SAW in the last chapter that there are many more farms and farm-workers on the plains than in the barren hills. Such has been the case since Saxon times, and Map **A** shows clearly that in the 18th century the better agricultural districts had the densest population. (*(1) Name four such districts.*) There is no such simple explanation of population Map **B**. Compare it carefully with the map on page 64. (*2*) What do you notice about the populous areas numbered 1 to 9? (*3*) Give each of them a suitable geographical name, e.g.: 6 = the Potteries.

The remarkable changes in the distribution of population shown on these maps took place because in the late 18th and 19th centuries Great Britain became a great manufacturing and mining country. The most important single cause was the application of steam power to industry: right down to the 1920s large-scale industry developed mainly on the coalfields of the Midlands and the North; and it was mainly in these areas that the phenomenal increase in Britain's population was concentrated.

We became the ' workshop of the world ', and British manufactures were bought by people in all continents. In return they sent us raw materials and cheap foodstuffs to feed our growing population. Faced with such competition, British farmers suffered a set-back from which there was no proper recovery until the Second World War.

There is no need today for a factory to be built on or even near a coalfield. (*(4) Why?*) In spite of this, the coalfield areas remain centres of the traditional **basic industries.** This is largely due to geographical inertia (*p. 80*).

(*5*) Name the chief industries in the areas marked 1 to 9 on Map B opposite. As well as those employed in these industries, even larger numbers work in the professions, in entertainment, commerce, transport and communications, building and public administration. Almost half the people of Britain are still packed into the great conurbations of the coalfield districts. (*6*) Where is the only other very large concentration of population?

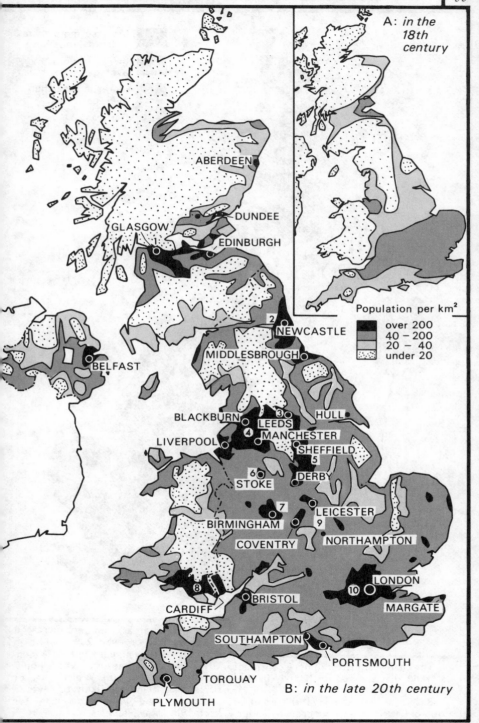

A: *in the 18th century*

Population per km²

- over 200
- 40 – 200
- 20 – 40
- under 20

ABERDEEN

DUNDEE

GLASGOW

EDINBURGH

NEWCASTLE

MIDDLESBROUGH

BELFAST

BLACKBURN

LEEDS

HULL

MANCHESTER

LIVERPOOL

SHEFFIELD

STOKE

DERBY

BIRMINGHAM

LEICESTER

COVENTRY

NORTHAMPTON

LONDON

BRISTOL

MARGATE

CARDIFF

SOUTHAMPTON

PORTSMOUTH

TORQUAY

PLYMOUTH

B: *in the late 20th century*

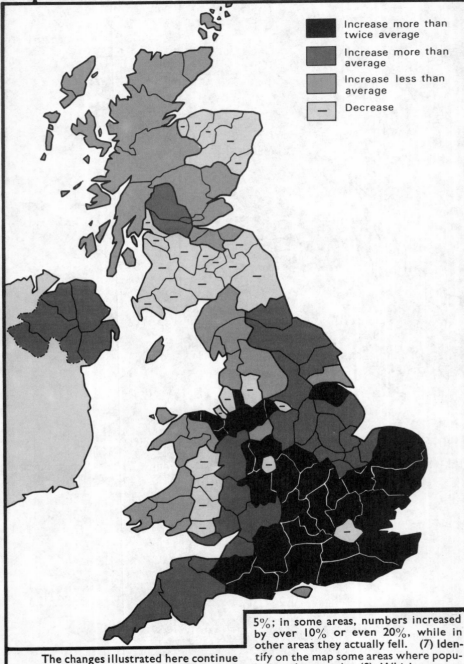

Increase more than twice average

Increase more than average

Increase less than average

Decrease

The changes illustrated here continue a trend that showed itself early in the 20th century.

Between 1961 and 1971 the population of Britain as a whole grew by about 5%; in some areas, numbers increased by over 10% or even 20%, while in other areas they actually fell. (7) Identify on the map some areas where population decreased. (8) Which part or parts of Scotland are likely to reveal a changed trend when the map for 1971–1981 is available?

The map opposite reflects the fact that industries and their work-people are tied less rigidly to the coalfields than formerly. In recent years a great many factories concerned with *new* industries have been built in the Midlands and the south—especially in and around London (*details, pp. 184 and 263*). Many of their work-people came from districts like Tyneside, South Wales and Lancashire, where industries like coalmining, shipbuilding and cotton manufacturing were especially affected by the 'depression' of the 1930s. Some such districts have never fully recovered from the unemployment and other economic dislocations of that time, for the old-established industries on which they depended have encountered more and more competition from newly-equipped factories in other countries.

To help these areas, of which South Wales, Central Scotland and Northern Ireland are good examples, the Government encourages manufacturers to start new factories and industries. Grants and loans are offered for the purchase of new plant, machinery and buildings, and removal expenses are paid for firms coming from elsewhere. These inducements have to some extent revived prosperity in many problem areas, but in spite of them industry and population continue to grow steadily along a so-called 'growth axis' between London and Liverpool which can be clearly seen in the map. The 'pull of the market' is now perhaps the strongest factor in the location of most new industries. (*9*) *Explain this statement, after a further look at the last two paragraphs.* This persistent trend has brought problems of congestion, high cost of land and shortage of labour inside the area of growth; while other areas remain in need of Government help even though their resources of land and labour are still under-employed. (See also pages 94–7, 106, 117 and 130–5.)

During the past century there has been a continued drift of population to the cities from certain country districts (see also page 70). (*10*) What are the reasons for this movement? (*11*) From the map make a list of the regions chiefly affected by this 'rural depopulation'. (*12*) Make a second list of regions in which the population increased by more than twice the average during 1961–1971, and suggest reasons based on earlier chapters. (*13*) Name some densely-populated urban areas where the population has recently decreased, and suggest reasons. (*14*) This map suggests that the overall distribution of Britain's population is becoming (*more/less?*) like that of the 18th century.

All cities were once towns, and most towns began as villages. Why are settlements, whatever their size, located where they are, instead of somewhere else? The answer usually has something to do with the transport of people or goods. This costs time and money; and for cheapness, any trade route follows as straight a course as possible. Few routes, however, go in a dead-straight line, and the exact paths followed by canal, road and even air routes are largely determined by the geography of the districts over which they pass. Where possible, land routes avoid wide rivers, marshes, mountains and lakes (see maps on pages 68, 89 and 226). Places at which such obstacles could be by-passed often became important **route centres**. A ' **nodal point** ' was thus formed, towards which travellers and goods converged, and the place often became a market town. (*15*) Write down as many names as you can of towns in the British Isles containing the words ' bridge ' and ' ford '; e.g. Waterford, Wadebridge.

When the obstacle could not be by-passed elsewhere, the route centre had great strategic value, and was often fortified. (*16*) Draw fully labelled sketch-maps to show which routes the following castles were built to defend (use the index and an atlas): Harlech; Stirling; Dover. (*17*) Which of these strategic positions most resembles that at Edinburgh (page 84); at Ludlow (page 100); and at Dublin (page 238)? (*18*) Where is the nearest castle to your school? Who built it, and when? (*19*) Draw a labelled sketch-map to show what your local castle was built to defend. (*20*) To what extent was the castle's site, i.e. the ground on which it was built, chosen so as to take advantage of any features on the ground which naturally aided defence? (*21*) Make notes, based on chapter 15 and on pages 278–81, to show how the growth of London illustrates all the points made above.

The location of many early towns in Britain was chosen with a view to defence, but in the course of this book we have seen that many other factors can play a part in influencing both the siting of a town and its subsequent growth. (*22*) Pair off correctly the ' heads ' and ' tails ' below.

Wantage		Lowest bridging point
Ely		Railway junction town
Gloucester		' Gap town '
Cork		Coalfield industrial town
Swindon	**?**	Spring-line settlement
Harlow		Dry site in former marsh
Wendover		Large natural harbour
Taunton		New planned town
Ebbw Vale		Old water-power industrial town
Stroud		Market town

(*23*) Can you suggest two suitable alternatives for each of the towns in your corrected pairs?

(*24*) Look closely at this photograph and explain why you think the village grew up at this particular place on the coast. (*25*) What clues does the photograph give about the probable occupations of the people (*a*) in the village and (*b*) inland? (*26*) Describe the site of the village and suggest, with reasons, a location for it within the British Isles.

This traffic island is near the centre of the ancient town of Hythe, in Kent. How many signs and other bits of so-called ' street furniture ' can you count on it? More travelling means more traffic. More traffic mean more noise, fumes and danger, as well a more clutter of this kind. What is the solution?

The rapid growth of industries and population during the 19th century created an enormous demand for houses in towns on the coalfields. In most mining and manufacturing districts row upon row of small, shoddy, ' back-to-back ' dwellings were built, often without gardens or proper sanitation. Mean, narrow streets were laid out around the factories and pit-heads, so that workpeople and their families were never free from smoke and grime. As the towns expanded, all who could afford to do so moved out to the fresher air and pleasanter surroundings in the suburbs. Meanwhile, houses in the centre of towns fell into decay, and were often let off into grossly overcrowded flats. Such was the origin of the sordid slums which still disfigure many of our larger industrial cities.

In the 1930s many of the worst slums were demolished, for their dirt, disease and overcrowded tenements were a menace to the health of the whole community. The displaced inhabitants were re-housed at public expense in special new estates, often on the outskirts of the town. It is in the suburbs, too, that most new private housing estates have been built, and we have noticed especially the spectacular growth of Greater London (*pp. 186–7*).

One of these old houses is built of Cornish
~nite and the other of Kentish timber and
ts, and there is no mistaking which is
ch. Each of them fits naturally and
asantly into its geographical setting. (27)
~at local building materials and styles
:inguish your home district from other
ts of Britain?

Compare these dwellings with those on an
~rage housing estate (below), whose drab
~formity has nothing local or lively about

it. They could be—and unfortunately are—
almost anywhere in Britain.

There are official bodies and private soci-
eties concerned to protect both town and
countryside against indiscriminate ' development'. (28) Find out all you can about the
functions of: the Civic Trust; the National
Trust; the National Parks Commission; the
Council for the Protection of Rural England;
the County Planning authorities; the Department of the Environment.

In their efforts to live in semi-rural surroundings the inhabitants
of our industrial towns and cities have created new problems, the
chief of which is the daily journey, tiring and increasingly expensive, between their residential ' dormitories ' and their place
of work. One attempt to solve this ' commuting ' problem and
to limit the further spread of huge conurbations, has been the
creation of ' New Towns ' with both jobs and homes planned in
detail before any building is started. It now seems unlikely that
any more New Towns will be established in the foreseeable future
(pp. 282–3.)

POPULATION OF BRIGHTON

Year	Population
1761	2 000
1786	3 600
1801	7 339
1811	12 012
1821	24 429
1831	40 634
1841	46 661
1851	65 573
1861	77 693
1871	90 011
1881	99 091
1891	115 873
1901	123 478
1911	131 237
1921	142 430
1931	147 427
1951	156 486
1961	162 757
1971	163 860

Many of the worst examples of unsightly building have occurred in coastal resorts and residential towns, even in places which rely mainly on their natural beauty to attract holiday-makers and tourists. The haphazard, unplanned growth of such towns as Brighton and Blackpool (population 148 000) accompanied the rapid growth in importance of the holiday and entertainment industries. This was made possible by the development of a network of cheap, fast, safe transport; railways in the 19th century and motor roads in the 20th. (29) Draw a graph to show the increase of population in Brighton between 1801–1971.

In the last few pages we have seen how the landscape of Britain is constantly being altered—not always for the better (see also page 159). The fortunes and importance of towns and cities change, as some industries expand and others decline. For instance, the population of Corby in Northamptonshire increased between 1931 and 1971 from 1596 persons to 47 940 as a result of the rapid growth of the iron and steel industry on the Jurassic escarpment.

 = 1 million male workers   **WORKING POPULATION IN GREAT BRITAIN: 1971**   1 million female workers =

The notes opposite show that important changes have taken place in the various occupations of the British population, as well as in its distribution. You will notice that the upper table is arranged so that the occupations are in the order of their numerical importance in 1841. (30) Make a similar list, arranging the same occupations in the order of their numerical importance in 1971. (31) Then make two sets of bar diagrams to represent each of the lists (i.e. that for 1841 in the book, and the one for 1971 you have written down). Shade the bars representing each occupation in a different colour. Your finished diagrams will show clearly the changing fortunes of the various occupations.

(32) From the notes, write down two lists: one containing the occupations which have decreased in importance, and the other showing those which have increased in importance during the past century. (33) The main causes of these changes are given in the lower table opposite. Can you spot the links?

(34) Using the last column in the table, make a 'body' diagram (similar to that above), dividing the working population into its various occupations. Work to the nearest half-million: e.g. let 14 'bodies' represent Manufacturing. Shade the 'bodies' representing each occupation in a different colour, and add a labelled key.

INDUSTRIAL DISTRIBUTION OF THE WORKING POPULATION 1841–1971

Occupations	Percentage Distribution 1841	Percentage Distribution 1971	Numbers (000s) 1971
Manufacturing (including gas, electricity and water supply)	32·0	28·7	7 139
Farming, forestry and fishing	23·0	2·7	670
Building and contracting	6·0	6·1	1 535
Trade (*shops, etc.*), commerce, insurance, banking and finance	5·7	15·6	3 853
Professions (teaching, nursing, etc.) and services (catering, hairdressing, etc.)	4·0	20·6	5 097
Technical and clerical workers in industry, including managers	3·5	10·2	2 533
Mining and quarrying	3·0	1·5	374
Transport and communications	2·9	6·3	1 570
Armed forces	0·8	2·3	574
Central and local government	0·6	5·6	1 390

The actual size of the working population more than doubled in this period. Another great change was the entry of large numbers of women into most types of employment. ' Manufacturing ' covers a very wide range of industries. Some declined in relative importance between 1841 and 1971 (e.g. textiles and shipbuilding), while others (e.g. printing and food-processing) expanded enormously. Some industries important in 1971 (e.g. motor vehicle and electrical engineering) were unknown in 1841.

SOME IMPORTANT CAUSES OF CHANGE IN THE RELATIVE IMPORTANCE OF OCCUPATIONS IN GREAT BRITAIN DURING THE PAST CENTURY

Increasing numbers of British farms have become fully mechanised.

Wages and salaries have tended to become more equal.

Permanent shops have very largely replaced weekly markets.

Britain has become largely dependent on imported foodstuffs.

Almost everyone can now read and write.

The number of business loans, and payments settled by cheque, has increased.

As the mass of people have become better off they could afford to spend more on comforts, luxuries and entertainment.

The past century has seen the development of : steamships, railways, motor vehicles and aeroplanes, as well as postal, telegraph and wireless services.

Both overseas and internal trade have greatly expanded.

The amount of business conducted by the State has expanded beyond recognition, especially in providing against unemployment, sickness and old age.

Cheap coal and iron mined in Britain formed the basis of our wealth in the 19th century. Today mining is increasingly mechanised and our iron ore resources are largely ' worked out '.

Modern warfare demands skilled professionals ready for instant action.

Modern factories produce more goods with fewer workers.

Health, education and welfare services have greatly expanded.

Electricity is sent to all parts of Britain through an elaborate network of transmission cables, commonly called the Grid. This photograph shows the Grid power-lines crossing the Thames at Dagenham. (35) Why are the cables held so high above the water?

Because power was lost in transmission, it was once necessary to build electricity generating stations near the district to be served. Because of the limited size of generators, no plant could produce more than a few hundred megawatts. A complete map of British power stations, including the smallest and oldest, is thus very similar to a population map. ((36) Why?)

Modern technology, with giant generators and a 400 000 volt 'supergrid' distributing the current over long distances, now allows power stations to be built in locations best suited to efficient operation. (37) Of the twelve very large coal-fired power stations marked on the map on pages 268–9, how many stand on or very close to the River Trent? (38) Suggest the advantages of their location in respect of (a) fuel, (b) cooling water, (c) transmission of power throughout the country.

(B) ENERGY RESOURCES

Within the past few years the word 'energy' has replaced the older term 'fuel and power'. In the 18th century the distinction was still clear. *Fuel* for (*e.g.*) heating, cooking or iron-smelting was wood or coal. *Power* was provided by wind, water or human and animal muscles.

The Industrial Revolution of the 18th–19th centuries was based on coal-fired steam engines. In Britain, its original forests almost cleared, 'King Coal' reigned unchallenged as the main fuel and thus as the major source of power for industry and for long-distance transport. By the mid-20th century, almost all factories and railways were electrified, and petroleum in some form was the chief power source for road, sea and air transport and for work on the land. Except for such direct uses as heating and cooking, it had become hard to distinguish between 'fuel' and 'power'.

With these changes came an enormous increase in the consumption of energy for all purposes. (39) *How many jobs can you think of that were done, as recently as 1900, by human or animal muscle*

*The world's first nuclear power station to gen-
ate current for commercial use was at Calder
all, in Cumberland (above). In 1957 it
gan to feed electricity into the Grid, making
itain the first country to use nuclear energy.*
*Very great quantities of water are needed to cool the machinery in power stations, so that
herever possible they are built beside large rivers or on the coast. Alternatively, concrete
oling towers (above, left) allow the same water to be used over and over again.*

power, and are now done by machines? Rising standards of living,
too, have demanded more energy for heating, cooling, cleaning,
cooking, and so on, in our homes and workplaces. Most of this
energy has been in the form of electricity.

Thus we often contrast the 19th century ' Age of Steam ' with
the 20th century ' Age of Electricity '. Yet electricity has to be
generated, and most British power stations still depend on coal-
fuelled, steam-driven generators (*diagram B overleaf*). This in
turn has meant that the electricity industry has become the
National Coal Board's chief customer, and is likely to remain so
for some years to come.

None the less there has been a revolutionary change in the
supply of fuel and power. From a ' one-fuel economy ' we have
moved to a ' five-fuel economy '. To coal we have added hydro-
electricity and petroleum and, since the 1950s, nuclear power and
natural gas. All of these are used for generating electricity, though
petroleum and gas are more widely used in transport and industry
respectively. From a 99% share of Britain's primary energy
consumption in 1913, coal has declined to a mere 38% (*C overleaf*). Factories use only a sixth, homes only a third of the

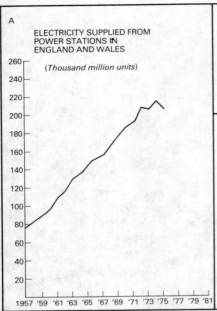

A

ELECTRICITY SUPPLIED FROM POWER STATIONS IN ENGLAND AND WALES

(Thousand million units)

260
240
220
200
180
160
140
120
100
80
60
40
20

1957 '59 '61 '63 '65 '67 '69 '71 '73 '75 '77 '79 '81

B

SOURCES OF ELECTRICAL POWER IN 1976 IN ENGLAND AND WALES (m's of tonnes coal equivalent*)

| COAL | | | | | | | OIL | NUCLEAR | HYD | G |

10 20 30 40 50 60 70 80 90 100 %

C

PRIMARY FUEL CONSUMPTION (GREAT BRITAIN)

(million tonnes coal equivalent *)

	1913	1966	1975	1985
	(actual figures)			(forecast)
Coal	186	175	121	130
Oil	2	112	133	188
Nuclear	—	7	11	25
Hydro	—	3	3	3
Natural gas	—	1	55	70
	188	298	323	416

* '1 tonne coal equivalent' is the amount of fuel quired to produce the heat of 1 tonne of c e.g. 1 tonne of oil = 1.7 tonne coal equivalent.

A and B illustrate some of the points already made. (40) Without quoting individual figures, state what graph A tells us. (41) State the approximate proportions of each power-source shown in diagram B. In 1971 the figures for the three main items were: coal 66%, oil 24%, nuclear 8%. Account for the subsequent changes.

Table C includes energy sources used directly in road transport, home cooking and heating, iron smelting and other ways as well as in the generation of electricity. The increase in total output of energy since 1913 is even greater than it appears, because modern appliances use fuel much more efficiently and so make each tonne go further.

(42) Convert C to a single graph as in A (with five separate curves) or to four bar diagrams as in B (one for each year quoted). Comment on and try to account for the main changes noted.

1913 quantities, and two other important markets for coal, the railways and the gas industry, now use none. Only power stations use more—over ten times more—than in 1913.

Coal lost its place partly because of rising costs as the more easily-mined seams became 'worked out'; partly because other sources of energy—oil, natural gas and electricity—became for a

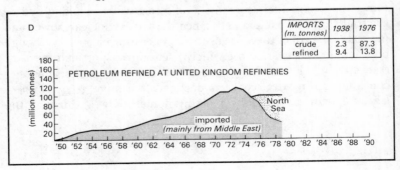

D

IMPORTS (m. tonnes)	1938	1976
crude	2.3	87.3
refined	9.4	13.8

PETROLEUM REFINED AT UNITED KINGDOM REFINERIES

(million tonnes)

180
160
140
120
100
80
60
40
20

imported (mainly from Middle East)

North Sea

'50 '52 '54 '56 '58 '60 '62 '64 '66 '68 '70 '72 '74 '76 '78 '80 '82 '84 '86 '88 '90

while abundantly and cheaply available. The number of mines fell from 948 in 1947, when the coal industry was nationalised ((*43*) *What does this mean?*) to 240 in 1976; and the work force of 704 000 miners in 1947 fell to 240 000 by 1976. In the same years, admittedly, the average output per mine, and per man, rose sharply as a result of mechanisation. None the less, the overall decline was marked. (*44*) *Using the table overleaf, draw bar diagrams to represent the output of coal in 1913, 1952 and 1976 from each major coalfield.*

For almost thirty years after the end of World War 2 the consumption of petroleum, as of electricity, doubled every eight or nine years. Diagram D shows the great changes that took place in the British oil-refining industry between 1949 and 1975. The main reasons for this vast expansion were:

(*i*) demand for all petroleum products rose enormously;

(*ii*) so many different refined products—aviation spirit, kerosene, diesel oil, lubricants, paraffin, etc.—are used today that it is cheaper to make them here from crude oil, imported in bulk, than to import them separately;

(*iii*) until about 1950 Britain's oil imports were mainly refined products from the U.S.A. Rising consumption in the U.S.A. left Britain, like the rest of Western Europe, dependent on the Middle East, which had most of the world's oil but very few refineries;

(*iv*) supplies of imported oil are liable to be cut off by disturbances in such regions as the Middle East;

(*v*) improved methods of refining leave many by-products which can be used in industries such as petro-chemicals and plastics.

Oil refining is now among Britain's major industries, and by the early 1970s crude and refined oil made up (*by weight*) about half of all our imports. Diagram D thus reminds us of the extraordinary stroke of good fortune which has presented Britain with a major supply of home-produced oil. Following the discovery in 1964 of natural gas in great quantities beneath the North Sea, oil and gas in even greater quantities were found further north from 1970 onwards. More gas and oil fields are still being located, and the map overleaf will probably be out of date before you see it. (*45*) *What details can you add to it, from your own knowledge of current events?*

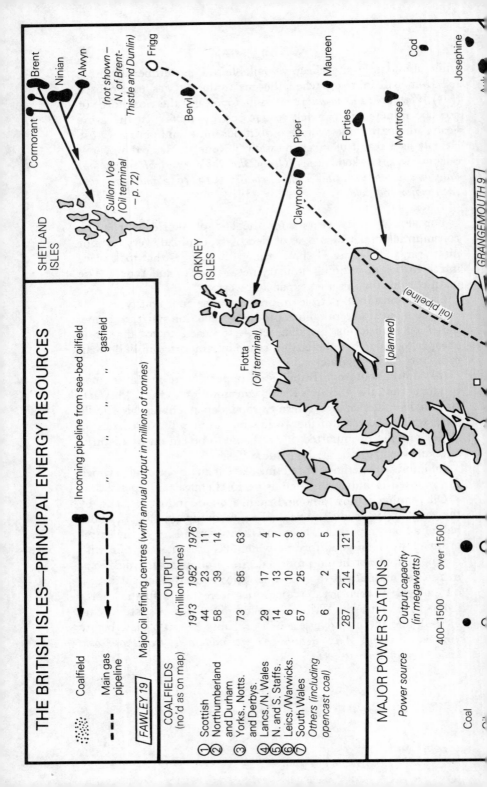

THE BRITISH ISLES—PRINCIPAL ENERGY RESOURCES

Coalfield

Incoming pipeline from sea-bed oilfield

Main gas pipeline

" " " " " gasfield

FAWLEY 19   Major oil refining centres (*with annual output in millions of tonnes*)

COALFIELDS (no'd as on map)	OUTPUT (million tonnes)		
	1913	1952	1976
① Scottish	44	23	11
② Northumberland and Durham	58	39	14
③ Yorks., Notts. and Derbys.	73	85	63
④ Lancs./N. Wales	29	17	4
⑤ N. and S. Staffs.	14	13	7
⑥ Leics./Warwicks.	6	10	9
⑦ South Wales	57	25	8
Others (including opencast coal)	6	2	5
	287	214	121

MAJOR POWER STATIONS

Power source   Output capacity (in megawatts)

400–1500   over 1500

Coal

Shetland Isles

Brent

Cormorant

Ninian

Alwyn

(not shown – N. of Brent – Thistle and Dunlin)

Sullom Voe (Oil terminal – p. 72)

Beryl

Frigg

Maureen

Cod

Josephine

Piper

Claymore

Forties

Montrose

Orkney Isles

Flotta (Oil terminal)

(oil pipeline)

(planned)

GRANGEMOUTH 9

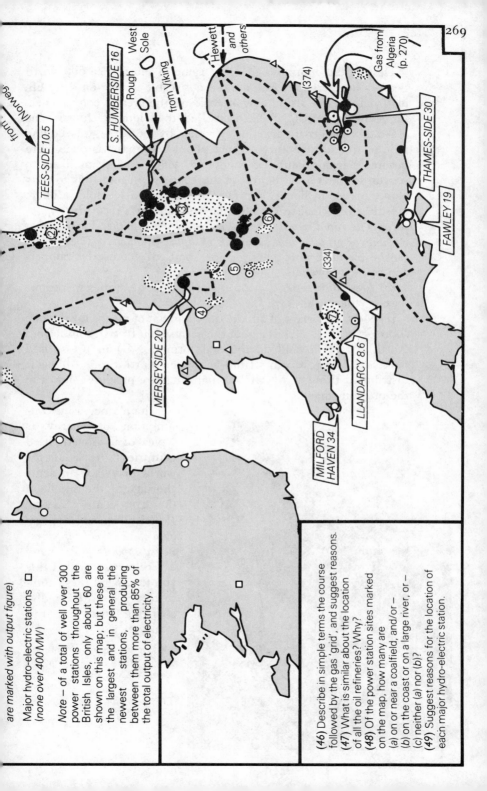

are marked with output figure)

Major hydro-electric stations ☐
(none over 400 MW)

Note – of a total of well over 300 power stations throughout the British Isles, only about 60 are shown on this map; but these are the largest and in general the newest stations, producing between them more than 85% of the total output of electricity.

from— (Norway)

TEES-SIDE 10.5

Rough   West
        Sole
S. HUMBERSIDE 16

from Viking

Hewett *and others*

(374)

Gas from Algeria (p. 270)

THAMES-SIDE 30

FAWLEY 19

MERSEYSIDE 20

MILFORD HAVEN 34

LLANDARCY 8.6

(334)

(46) Describe in simple terms the course followed by the gas 'grid', and suggest reasons.
(47) What is similar about the location of all the oil refineries? Why?
(48) Of the power station sites marked on the map, how many are
(a) on or near a coalfield, and/or –
(b) on the coast or on a large river, or –
(c) neither (a) nor (b)?
(49) Suggest reasons for the location of each major hydro-electric station.

Despite the enormous financial value of North Sea oil, North Sea gas has perhaps made a more noticeable impact on daily life in Britain. The rising cost of coal and the greater convenience, for some purposes, of electricity led to a long decline in the old-established coal-gas industry which, by the 1950s, was a dying relic of the 19th century. Since about 1960 gas has once more become a major source of energy (*p. 266C*) and seems likely to remain so for at least the rest of this century. From 1965 to 1975 alone, its use in homes and factories increased fourfold. There are three principal reasons for the gas industry's recovery:

(*a*) *The rapid world increase in petroleum refining, leaving large amounts of propane and butane as cheap by-products.* These gases can be blended with other gas, or ' bottled ' for use by campers and country-dwellers.

(*b*) *Fresh discoveries of natural gas* in great quantities in various parts of the world.

(*c*) Most important of all, the development of means for the *safe transport of gas*. About 2% of the gas used in Britain comes from North Africa in chilled liquefied form (LNG) in pressurised tankers. The rest comes from below the North Sea. All gas is distributed through a ' grid ' of high-pressure pipelines shown in the previous map.

Britain's increasing consumption of energy is typical of all industrialised countries. As a result the world may face an acute shortage by the middle of the 21st century*. (*50*) *Will Britain be able to deal with the problems mentioned opposite*? Press and TV references will enable you to expand these notes to keep pace with rapidly-changing ideas on this vital topic.

* For a general summary of world energy resources, see Book 3 Chap. 9.

Laying a natural gas pipeline for Burton-on-Trent.

Far out in the North Sea a 'blow-out' on this offshore oil rig threatens a disastrous fire which the stand-by vessel's hoses have so far prevented. Scenes like this help to explain the rising cost of oil (*see notes*).

WHAT FUTURE FOR BRITAIN'S ENERGY SUPPLIES?

COAL Britain has enough for 300 years (at current rates of mining) and large new discoveries are still being made. Coal can be converted into substitutes for oil and gas; but can it be mined and processed for this purpose at an acceptable cost? This is still not certain. (*Note: despite a 50% decline since 1913, British coal output is still the world's fifth largest.*)

OIL and NATURAL GAS By 1980 Britain should be self-sufficient in oil as well as in gas. By the 1990s, however—and certainly early in the 21st century—output will be past its peak. Big new offshore discoveries are possible, but these will increasingly be in deeper, stormier seas, entailing higher costs of production.

HYDRO-ELECTRICITY No existing or possible future plant in Britain can generate as much electricity as a modern 'thermal' power station.

Tidal barrages are a possible, but expensive, source of hydro-electricity.

NUCLEAR POWER Nuclear fission plants now in service produce electricity more cheaply than coal, oil and gas-burning stations. One school of thought sees nuclear power stations as the only way to close the 'energy gap' which will widen as demand rises and **fossil fuels** become scarce and expensive. Others claim that the risks associated with dependence on nuclear fission are unacceptable. A safer process—nuclear fusion—seems unlikely to be in use before AD 2050 at earliest.

ALTERNATIVE SOURCES OF ENERGY Electricity can be generated by using waves, wind, sunlight and the heat of the Earth's interior. The problem is to produce power at reasonable cost. Researchers think that 5% of Britain's energy needs might be met in one or other of these ways by the end of the century.

Most people in their own homes now enjoy a standard of comfort that would have seemed luxury to almost everyone a century ago. Most people on holiday can now afford to go further afield than their grandparents ever dreamed of going. All this is welcome progress. Yet two kinds of progress ma[y] conflict with each other.

(51) What modern problem is illustrate[d] in this view of a giant aircraft landing a[t] London Airport?

The increases mentioned so frequently in the past few pages have helped to bring a far higher standard of living to a growing population. Rather suddenly in the 1970s we have begun to realise that the price of ' progress ' may in some ways be high.

To take just one example, it is hard to imagine life without plastics and paper; but **effluent** from plastics and paper factories has been allowed for too long to poison the rivers from which, increasingly, we need to draw drinking water.

Oil slicks on holiday beaches; the need for copper and gold mining in Snowdonia National Park, or for afforestation in the Lake District; traffic jams in towns; traffic accidents on motorways; the best location for London's third airport—these are a few of the problems and controversies that arose in the 1960s as a direct result of our ever-growing demands on the great, but still limited, resources of our environment. They are problems that remained unsolved in the 1970s.

(52) Find out the facts behind some of these problems with the arguments put forward in favour of various solutions. (53) Which solution do you favour in each case?

A trading nation needs an efficient transport system. Britain's roads, railways and ports have become outdated, and we are having to spend large sums of money to remove this handicap.

The ' container ' system became a feature of world trade during the 1960s. Here is a container ship being loaded at Southampton. (54) What are the advantages of using containers? (55) How is the system applied to railways? (56)—to road transport?

(C) TRADE AND TRANSPORT

Britain was the first country in the world to develop large manufacturing industries, and for long we supplied other countries with industrial goods which they were unable to make for themselves. At the same time, as we have seen, our rapidly growing population came to rely on other countries to supply us with one-half of all the food we eat, as well as many vital raw materials needed for our manufactures. All through the 19th century Britain was the world's largest trader, and for most of it the world's largest industrial producer. Both these positions are now held by the U.S.A., but Britain is still a leading industrial nation.

Industrial activity in Britain has increased rapidly in recent years: for example, in 1970 it was over twice the figure for 1940. The most striking example has been in chemical manufacturing, based on the growth, from virtually nothing, of an immense oil-refining industry. Even more important has been the expansion in some forms of engineering—e.g. the manufacture of aircraft,

273

motor vehicles, machinery, electrical goods, precision instruments and radio equipment.

Diagram B opposite shows what British industry produces now. (57) List the various items in order of value. Next, write a general account of our overseas trade from diagrams A and C. (58) Now look at the more detailed notes on page 277, and state what effects each of the following might have on our ability to ' pay our way ' in overseas trading:—

(i) A big fall in meat prices.
(ii) A ban on smoking in British cinemas, theatres and trains.
(iii) A ban on the import of vehicles and whisky into the U.S.A.
(iv) The growth of a large-scale chemical industry in Australia.

Using pages 275–7 for information:—

(59) Draw a bar diagram showing to scale Britain's chief imports. Suggest the main sources of the goods in each case.

(60) Draw a similar diagram of Britain's chief exports, and suggest the main destination of the goods in each case.

Until very recently a large proportion of Britain's trade was with Commonwealth countries. After 1950, however, a new and fast-growing trend became apparent, i.e. a great increase in trade with Europe. This trend accelerated sharply after Britain's entry into the Common Market in 1973. There has been a steady decline in the proportion of our trade which flows to and from the Commonwealth, though the total volume of that trade remains large for the following reasons:—

(a) Commonwealth countries produce many raw materials and foodstuffs which we greatly desire but cannot produce for ourselves. This may be because our climate is unsuitable, e.g. in the case of tea, cane sugar, coffee and rubber; or because we lack sufficient agricultural land, e.g. to produce enough wool, dairy produce and cereals; or because we lack minerals, such as tin, copper, petroleum and bauxite.

(b) Although manufacturing industries are developing quickly in most Commonwealth countries, there is still a big demand for many types of British manufacture.

(c) Our commercial links with the Commonwealth are still (though to a decreasing extent) strengthened by the mere habit of a century and more of ' Empire ' trade, as well as by ties of language and loyalty to the Crown.

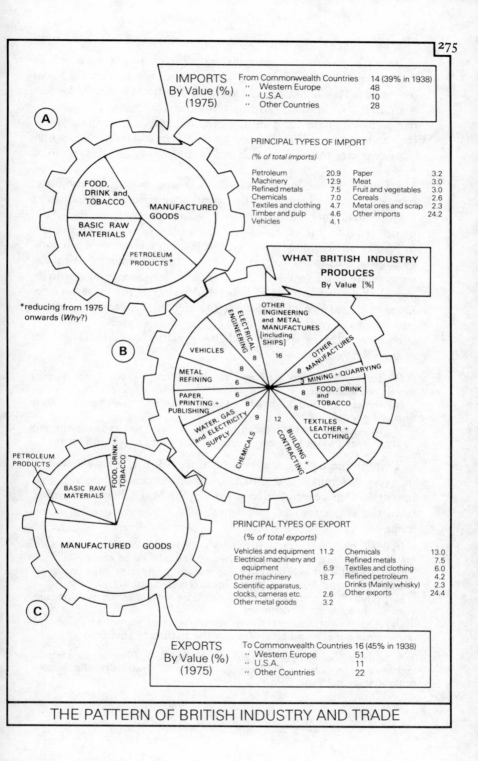

IMPORTS
By Value (%)
(1975)

From Commonwealth Countries	14 (39% in 1938)
Western Europe	48
U.S.A.	10
Other Countries	28

A

FOOD, DRINK and TOBACCO

MANUFACTURED GOODS

BASIC RAW MATERIALS

PETROLEUM PRODUCTS *

PRINCIPAL TYPES OF IMPORT

(% of total imports)

Petroleum	20.9	Paper	3.2
Machinery	12.9	Meat	3.0
Refined metals	7.5	Fruit and vegetables	3.0
Chemicals	7.0	Cereals	2.6
Textiles and clothing	4.7	Metal ores and scrap	2.3
Timber and pulp	4.6	Other imports	24.2
Vehicles	4.1		

WHAT BRITISH INDUSTRY PRODUCES
By Value [%]

*reducing from 1975 onwards (*Why?*)

B

ELECTRICAL ENGINEERING

OTHER ENGINEERING and METAL MANUFACTURES [including SHIPS] 16

VEHICLES 8

OTHER MANUFACTURES 8

METAL REFINING 6

3 MINING + QUARRYING

PAPER, PRINTING + PUBLISHING 6

FOOD, DRINK and TOBACCO 8

WATER, GAS and ELECTRICITY SUPPLY 8

CHEMICALS 9

BUILDING + CONTRACTING 12

TEXTILES LEATHER + CLOTHING 8

C

PETROLEUM PRODUCTS

FOOD, DRINK + TOBACCO

BASIC RAW MATERIALS

MANUFACTURED GOODS

PRINCIPAL TYPES OF EXPORT

(% of total exports)

Vehicles and equipment	11.2	Chemicals	13.0
Electrical machinery and equipment	6.9	Refined metals	7.5
		Textiles and clothing	6.0
Other machinery	18.7	Refined petroleum	4.2
Scientific apparatus, clocks, cameras etc.	2.6	Drinks (Mainly whisky)	2.3
Other metal goods	3.2	Other exports	24.4

EXPORTS
By Value (%)
(1975)

To Commonwealth Countries	16 (45% in 1938)
Western Europe	51
U.S.A.	11
Other Countries	22

THE PATTERN OF BRITISH INDUSTRY AND TRADE

Britain's export trade is constantly in the news; so much so that we might well think most of the goods we produce are sent overseas. This is not so. In fact, foreign trade accounts for less than 10% of our total output. Yet this fraction is still vital to our well-being, for most of our raw materials and half our food have to be imported; and imports have to be paid for with exports. These exports are mainly *goods* such as those listed on the previous page.

A useful part of our import bill, however, is paid in **invisible exports**. When Britain led the world in commerce and industry, London was the centre to which businessmen throughout the world came to insure buildings and cargoes, to borrow money, to hire ships or to buy the advice of leading engineers, surveyors, geologists, doctors, lawyers and other technical experts. All these services are still in great demand, for the less-developed countries are all short of this kind of help; and payment for these services helps our ' balance of trade '. ((*61*) *Explain this term.*) Britain provides 11% of the world total of ' invisible exports '.

By contrast our ' visible ' trade has greatly changed. The 19th century system was:—

(*a*) to import cheap food and raw materials, mostly from under-developed countries;
(*b*) to feed the workers while they turned the raw materials into finished goods;
(*c*) to export manufactured goods, largely to underdeveloped countries, at a price high enough to pay for more of (*a*); and so on . . .

This simple pattern has largely broken down, for we now live in a very different world. Britain's trade reflects the changes that have taken place especially since 1950. (*62*) Choose examples from the opposite page to illustrate the following general statements:—

(*i*) Half of world trade (by weight) consists of petroleum sold mainly by underdeveloped countries to industrialised countries.*
(*ii*) The bulk of world trade (by value) comprises manufactured goods exchanged between highly industrialised countries.
(*iii*) The 19th century pattern can still be seen in much of our trade with Commonwealth countries.
(*iv*) The newer pattern shows clearly in the growing volume of trade with Europe and also—(*v*)—in the nature of that trade.

* By the 1980s we may have enough North Sea oil for our own needs, and perhaps a surplus for export. This will drastically change our present trade pattern.

BRITAIN'S PRINCIPAL TRADING PARTNERS

The countries and regions listed below account for 86% of Britain's foreign trade. The full details of our imports (*from*) and exports (*to*) every country in the list would include a bewildering variety of goods; only the main types of commodity, or particular commodities of special importance, are named here.

To save space, some items are represented by bold initials, thus: **P** = petroleum (**CP** = crude, **RP** = refined), **C** = chemicals, **T** = textiles, **M** = machinery and metal goods, **E** = electrical machinery and equipment, **V** = motor vehicles.

(*63*) Mark on a world map the facts given below, with labels proportionate in size to the amount of trade with Britain.

(*64*) Wherever you can, give geographical reasons why particular countries export certain commodities to Britain.

(*65*) Where broad areas (e.g. Latin America) are named, suggest individual countries supplying particular commodities, with reasons..

N.B. These details are correct for the mid-1970s. See, however, the footnote opposite.

UNITED STATES 12½%
Imports: **C, M**, steel, aircraft maize, wheat, tobacco
Exports: whisky, cars, aircraft, woollens

SCANDINAVIA (inc. FINLAND) 11%
Imports: meat, dairy produce, fish, timber, wood pulp, paper, steel, aluminium, **M, E, V**
Exports: **RP, C, T, M, E, V,** steel

'BENELUX' countries 7½%
Imports: meat, dairy produce, **RP, C, M, E, V**, steel, fruit, vegetables
Exports: **C, M, E, V**

WEST GERMANY 6%
Imports: **C, T, M, E, V**
Exports: **C, M, E, V**, furs, whisky, refined metals

CANADA 5½%
Imports: metals and ores, timber, pulp, paper, wheat
Exports: much as to U.S.A.

IRISH REPUBLIC 4%
Imports: meat, dairy produce, textiles, clothing
Exports: **C, T, M, E, V**, textiles, scientific instruments

FRANCE 4%
Imports: **C, M, E, V**, steel, wines, maize, wheat, meat, fruit, vegetables
Exports: **C, M, E, V**

AUSTRALIA 3½%
Imports: metals and ores, fruit, sugar, wheat, meat, wool, dairy produce
Exports: **C, T, M, E, V**, books

PERSIAN GULF STATES 3½%
Imports: **CP, RP**
Exports: **C, T, M, E, V**, cigarettes, aircraft

LATIN AMERICA 3½%
Imports: iron and copper ores, **CP, RP**, meat, fruit, coffee
Exports: **C, M, E, V**

SOUTH AFRICA 3%
Imports: fruit, maize, wool, gold, copper
Exports: **C, T, M, E, V**

ITALY 3%
Imports: fruit, vegetables, **RP**, **T, M, E, V**
Exports: **M, E, V**, metals, whisky, furs

SPAIN and PORTUGAL 2½%
Imports: fruit, wine, **T**, clothing
Exports: **M**, steel

WEST AFRICA 2%
Imports: cacao, lumber, **CP**, aluminium
Exports: **M, E, V**

SWITZERLAND 2%
Imports: **C, T, M**, precision engineering
Exports: **C, T, M, E, V**, clothing

U.S.S.R. 2%
Imports: wheat, timber, metals
Exports: **C, T, M, E, V**

JAPAN 1½%
Imports: **C, T, M, E, V**, fish
Exports: **C, M, E, V**, woollens

INDIA and PAKISTAN 1½%
Imports: tea, jute, cotton goods
Exports: **M, E, V**, steel

NEW ZEALAND 1½%
Imports: meat, wool, dairy produce
Exports: **C, T, M, E, V**

MALAYSIA and SINGAPORE 1¼%
Imports: rubber, lumber, palm oil
Exports: **M, E, V**

HONG KONG 1¼%
Imports: **T**, clothing
Exports: **M, E, V**

EAST AFRICA 1%
Imports: coffee, diamonds
Exports: **M, E, V**

AUSTRIA 1%
Imports: **T**, clothing
Exports: **T, M, E, V**

POLAND 1%
Imports: meat, timber
Exports: metals

ZAMBIA 1%
Imports: copper
Exports: **M, E, V**

Sugar from the West Indies; tea from India; butter and bacon from Denmark and bread made with Canadian flour . . . every day we eat a great variety of foods, but rarely give a thought to the vast journeys—by road, rail, canal, ship and sometimes air—involved in bringing them to our tables. This trade has been built up through the centuries, and has been made possible only by a steady growth in communications. Until Elizabethan times, people in Britain relied very largely on local resources for their food and clothing, and for the materials with which to make their dwellings and few possessions. The few people who travelled long distances, such as pilgrims and pedlars, either went by sea, or by boat on the navigable rivers, or else they followed ancient tracks (*opposite*) or the remains of Roman roads. (*66*) Why did most prehistoric tracks follow hill ridges? (*67*) Why did most Roman roads converge on London? (See map, page 179.) (*68*) Some Roman roads are followed by modern highways: are there any in your district?

Because they were little needed, medieval roads were often so neglected that in wet weather they became impassable. Routes across clay districts such as the Weald and the Vale of Oxford were particularly bad: in 1499, for instance, a man was drowned in a pool over 2 m deep on a main Buckinghamshire highway.

The reasons which led to rapid development in the late 18th century are mentioned opposite. Equally rapidly, however, the network of canals was outdated, even before it was complete, by the newly-invented railways.

The main function of the first railways (1825–1830) was to carry coal to the coast for export. ((*69*) Which coalfields were the following early tracks built to serve:—*Stockton and Darlington* (*see map, page* 112); *Glasgow and Garnkirk* (*page* 79); *Llanelli and Taff Vale Railways?*) Soon afterwards, railways designed to carry passengers and goods traffic spread throughout lowland Britain ((*70*) *Why there?*), thus facilitating the movement of population and the growth of industry referred to on pages 254 and 257. Some of the most important results of railway building were:—

(*a*) Industry and population grew quickly around many railway junctions and termini, and later alongside many main lines. ((*71*) Explain the reasons for this growth and name two important railway junction towns or termini.)

(*b*) Railways made possible the large-scale movement of goods in densely populated districts, and fast trains enabled people to

(Above) Part of the Icknield Way, a pre-storic track along the chalk hills from ...alisbury Plain to the Wash. Except during ...e period of Roman occupation, few of ...ritain's roads were any better than this until ...te in the 18th century, when the engineers ...elford and Macadam pioneered new methods ... road construction and surfacing.

Meanwhile, from the 15th century on-...ards, the ocean voyages of discovery had led ... a rapid growth of trade with the Americas, ...rica, India and the Far East. Later, in the ...th century, industry began to expand on the ...itish coalfields (p. 254) and the need to transport goods between inland districts and the ports greatly increased.

In the absence of good roads, this problem was first tackled by building canals (below). The part played by canals in the development of industry in the English Midlands, and their later decline in the face of competition from railways and new roads, was mentioned on pages 149–50. Many canals, like this one, have been drained and abandoned. Others are now used only for pleasure boating, angling and water supply. Very few British canals still carry any great volume of traffic.

Coast Town	Main Activity Originally Stimulated by Rail Link
Immingham Southampton Grimsby Middlesbrough Folkestone Harwich	Landing fish and despatching it to market Smelting iron and steel Catering for holiday-makers Handling continental passenger traffic Exporting coal Handling ocean passenger traffic

(72) Suggest two alternative coast towns mainly concerned with each of the activities mentioned above.

live far away from their places of work. Thus railways have encouraged the growth of conurbations and 'suburban sprawl'. (See also pages 186 and 260.)

(c) Certain coast towns developed their main activity entirely as a result of railway links. (73) Sort out the lists above into correct pairs.

(74) Using your atlas, draw a sketch map of Britain's main railway lines. Compare it with the map on page 179. Is our railway system a result, or a cause, of London's importance?

After drawing trade away from the canals, the railways formed Britain's chief means of inland transport for about a century. Over the past fifty years, however, they have been increasingly challenged by the enormous development of road traffic. Many towns which grew with the coming of railways have now lost their rail link, and many former branch lines now resemble the canal shown in the previous photograph. (75) Have any railway lines been closed in your district? (76) What were the reasons, and (77) what have been the results of closure? (78) What are some of the problems created by our dependence on road transport? (79) What major changes in everyday methods of transport and communications do you expect to see in your lifetime?

In addition to a complex network of roads, railways, canals, pipelines and airlines, Britain (like every other modern state) has a **telecommunications system** which plays a rapidly-growing part in our business and social life. (80) Construct graphs or diagrams to illustrate the following figures, and write a comment on the trends they reveal.

Year	1960	1965	1970	1975
Letters posted (thousand million)	10·2	11·2	11·4	10·9
Telephone calls (thousand million)	4·3	6·3	9·6	15·8

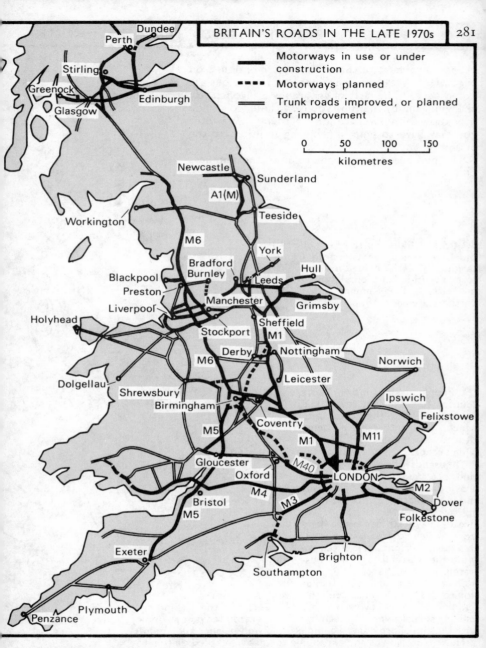

Motorways in use or under construction

Motorways planned

Trunk roads improved, or planned for improvement

0   50   100   150
kilometres

Over 90% of passenger travel and over 70% of goods transport is now by road.

(81) 'Convenience' and 'flexibility' are often said to be the great advantages of road transport over rail. Explain these terms, with examples. What are some of the counter-arguments in favour of rail transport?

The continuing rise in traffic has always outstripped the government's efforts to build new roads, but by the end of the 1960s a 'skeleton' motorway system was was in use.

(82) Draw a tall, thin capital H to represent the 'skeleton' referred to, and use it as the basis of a diagram to show how the motorways serve the main ports and centres of industry.

Some problems of the 1970s were mentioned on page 271. A few more problems and questions are noted here. The facts of geography are constantly changing, and by the time you read these words you may know the answers to some of them. As a voter you may have to help in choosing solutions to the others.

The numbered paragraphs refer to the numbers on the map.

① Both Britain and the Republic of Ireland have joined the European Common Market. Will this bring the two countries closer together, and possibly help to end the bitter disagreements that divide Northern Ireland?

② People are still moving from country to town, yet fewer and fewer of us live in the city centres, which are increasingly being cleared for shops, offices and wider roads.

Thousands of hectares of farm land are taken every year to provide new houses, schools, roads and work-places for people moving out from the cities. As we already have far too little land from which to feed ourselves, this loss is serious. How can we avoid the wasteful 'sprawl' illustrated on page 186?

Since 1945 an attempted solution to this problem has been the building of New Towns (pp. 80, 113, 261) and the expanding of existing small towns. In each case there is careful advance planning, with the aim of providing a complete and well-balanced community.

These New Towns and expanding towns are marked on the map. (83) Which New Town is nearest to your home? Have its planners achieved the aim mentioned above?

Britain's population grew steadily from 38 270 000 in 1901 to 55 745 000 in 1971. In the 1970s the rate of growth slowed rather suddenly, and by 1975 the total was actually falling. Partly as a result, the programme for 'New' and expanding towns was cut quite drastically in 1977 and government aid was directed more to the renewal of old city centres. (84) Suggest detailed reasons for this change of policy.

▲ "New Towns"
● "Expanding" tow

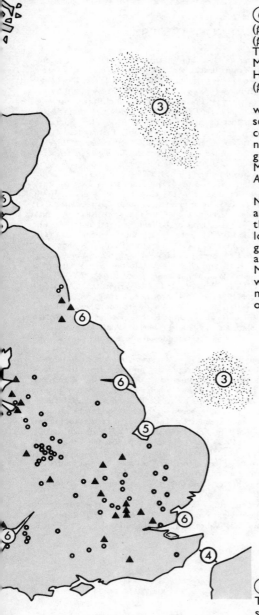

6 (85) Identify on the map Bantry Bay (p. 235), Milford Haven (p. 96), Severnside (p. 97), the Firths of Clyde (pp. 81–2), of Tay (p. 85) and of Forth (pp. 83–4), the Moray Firth (p. 71), Teesside (p. 114), Humberside (p. 141), the Maplin Sands (p. 177).

In the near future some of these deep-water inlets may be developed on a large scale as 'tidewater' industrial sites, accessible to the very large vessels which now carry the bulk of the world's cargoes. Which will be the first of the MIDAs (Marine Industrial Development Areas)?

The English Channel and southern North Sea are increasingly overcrowded and dangerous for shipping. It may be that the British Isles will become an unloading and transhipping point for bulk goods on their way to Northern Europe and even to the USSR. Bantry Bay and Milford Haven are already used in this way for oil. The 'Oceanspan' scheme mentioned on page 82 is another aspect of the same idea.

5 The demand for water for homes, factories and farms rises steadily. Already there are shortages, especially in the South-east. ((86) Why there? —give TWO reasons.) Every new reservoir 'drowns' land that can ill be spared. Several estuaries ((87) Identify and name those numbered on the map) are being studied with a view to turning them into huge freshwater lakes. Which, if any, will be the first? (See also pages 99 and 106). The dams (**barrages**) enclosing these lakes might also be used to generate hydro-electricity (p. 271).

4 Plans for the long-discussed Channel Tunnel project reached an advanced stage during the 1970s, but were once again shelved. Will they be revived in the 1980s? Can you suggest why many experts think that this might lead to the building of more factories, offices and houses in the south-east? Would this be a good or bad thing? (Reread pp. 183 and 257 before you decide.)

3 North Sea gas was found in the early '60s, and North Sea oil in the early 1970s. There is already enough to make a really substantial contribution to Britain's power supplies for some years to come. How much more is waiting to be discovered?

ACKNOWLEDGMENTS

We would like to thank the following for their kind permission to reproduce photographs used in this book (page numbers in brackets): Aerofilms (33, 34, 75, 76, 101, 105, 144, 154, 180, 185, 186, 204, 228–9, 241, 259); Agent General for Queensland (11); Eric Ager (153 top right); The Air Ministry (55, 59); Architectural Press photo by Ian Nairn (260); Barnaby's Picture Library (174); Beringer & Pampaluchi (35); *Birmingham Post* (151 right); G. Douglas Boulton (142); *Bradford Telegraph & Argus* (121 top, 123); British Aluminium Co. Ltd. (71); British Insulated Callender's Cables Ltd. (262); British Leyland (153 bottom); British Petroleum Co. Ltd. (97); British Rail (126); British Transport Docks Board (272); The Controller of H.M. Stationery Office (9, 12, 24, 28, 110, 202, 219); Valentine, Dundee (40); *Dunstable Borough Gazette* (171 top); Exclusive News Agency Ltd. (15, 16, 32, 50); *The Farmer & Stockbreeder* (90); Firth Vickers Stainless Steels (124); Fox Photos (38, 58); *Glasgow Herald & Evening Times* (67, 80, 81); Gloucestershire Newspapers Ltd. (221); Ham River Grit Co. (171 bottom); I.C.I. (115); Irish Tourist Board (234, 236); Chard Jenkins of Studio K (153 top left); W. T. Jones (279); Dr. J. K. S. St Joseph (39); Jute Industries Ltd. (85); *The Kent Messenger* (10, 194, 261); *Manchester Evening News* (133, 135); Elsam Mann & Cooper (128); Adolf Morath & The British Iron and Steel Federation (158); The Nevada State Highway Department (138); N. Ireland Development Council (233); Press Association Ltd. (230); *The Scotsman* (87); J. C. D. Smith (210 top, 218, 279 top); Stewarts & Lloyds Ltd. (270); Strelitz Ltd. (231); Studio Five (36); *The Times* (212); John Topham Ltd. (243, 250, 252); U.K. Atomic Energy Authority (263); Washington Development Corp. (113); Messrs. Josiah Wedgwood & Sons Ltd. (151); *The Western Morning News* (22, 210 bottom, 215); C. H. Wood (27); G. N. Wright (57); Aerofilms (14, 98); J. H. Lowry (23, 30); Douglas Scott (42); *Yorkshire Post* (60); *The Financial Times* (66, 182); British Steel Corporation and Mike Woodward (95); Swan Hunter Shipbuilders Ltd. (116); C. H. Wood Ltd. (121 bottom); Richard Sadler (149); British Steel Corporation (159); Delabole Slate Co. Ltd. and Charles Woolf (215 bottom); Press Association Ltd. (271). Every effort has been made, unsuccessfully, to trace the copyright owners of the remaining photographs.

We would also like to thank Ordnance Survey (*The Position of Helston*, p. 40 [Crown Copyright Reserved]); Macmillan & Co. Ltd. (2 on pp. 255 and 261, from their book *Elementary Economics* by Harvey); J. M. Dent & Sons Ltd. (3 on p. 192, based on those in their book *British Isles* by Thomas Pickles); Victor Gollancz Ltd. (1 at foot of p. 201, from their book *The Scenery of England and Wales* by A. E. Trueman); The Meteorological Office (the weather maps on pp. 59 and 61, based on those issued by the Director General, Meteorological Office); Longmans, Green & Co. Ltd. (1 on p. 31, 2 at foot of pp. 184 and 185, 1 on p. 204 and 1 on p. 256, all modified from their book *The British Isles*, by Stamp and Beaver); University Tutorial Press Ltd. (1 at foot of p. 191, from their book *The British Isles*, by Preece and Wood); The Clarendon Press (*British Rainfall*, on p. 62, based on that in the *Oxford School Atlas*); The Cambridge University Press (1 at foot of p. 206, from *An Historical Geography of England before A.D. 1800*, by H. C. Darby); The Geographical Association and A. H. Shorter (1 at foot of p. 216, from Dr. Shorter's article in *Geography*, Vol. 34, 1954).

Thanks are also due to Methuen & Co. Ltd. (1 extract from H. V. Morton's *In Search of England*); B. T. Batsford Ltd. (3 extracts from Black: *The Beauty of Britain*); and The Brewers' Society (3 extracts from Festival of Britain publications).

INDEX

288